# Science & Society

# Science & Society: Scientific Thought and Education for the 21st Century

Peter A. Daempfle, PhD
State University of New York – Delhi

JONES & BARTLETT
LEARNING

*World Headquarters*
Jones & Bartlett Learning
5 Wall Street
Burlington, MA 01803
978-443-5000
info@jblearning.com
www.jblearning.com

Jones & Bartlett Learning books and products are available through most bookstores and online booksellers. To contact Jones & Bartlett Learning directly, call 800-832-0034, fax 978-443-8000, or visit our website, www.jblearning.com.

**Production Credits**
Publisher: Kevin Sullivan
Senior Acquisitions Editor: Erin O'Connor
Editorial Assistant: Rachel Isaacs
Production Assistant: Alex Schab
Senior Marketing Manager: Andrea DeFronzo
V.P., Manufacturing and Inventory Control: Therese Connell
Composition: CAE Solutions Corp.
Cover Design: Scott Moden
Rights & Photo Research Assistant: Ashley Dos Santos
Cover Image: © Hemera/Thinkstock (Top Image) and © Pan Xunbin/ShutterStock, Inc. (Bottom Image)
Printing and Binding:McNaughton & Gunn
Cover Printing: McNaughton & Gunn

**Library of Congress Cataloging-in-Publication Data**

Daempfle, Peter, author.
  Science & society : scientific thought and education for the 21st century / Peter Daempfle.
    pages cm
  Includes index.
  ISBN 978-1-4496-8502-7 (alk. paper)
1. Science—Social aspects. 2. Science—Study and teaching. 3. Scientists—Psychology. I. Title.
II. Title: Science and society.
  Q175.5.D34 2013
  303.48'3—dc23

                                                                    2012024206

6048

Printed in the United States of America
24 23 22 21 20   10 9 8 7 6 5 4 3 2

*For my wife, Amy*
*For my children, Justina and Konrad*
*For my father, Tobias*

# Brief Contents

# Contents

# Preface

The purpose of *Science & Society* is to improve science literacy and advance the importance of scientific thinking. This book grew out of twenty years of science teaching with incoming students in need of a better foundation in both scientific concepts and how science is conceived. In my teaching, I felt an increasing need for a textbook that provided a comprehensive base for readers entering into a science discipline. *Science & Society* provides a unique approach to science by addressing it as a process and grounding the knowledge within the many facets that make it an exciting area of study.

The book shows the reader how to think like a scientist. It defines what science is and how it can be misused. Throughout the book, provocative science examples are provided that guide the reader to consider facts more critically. The tools to question authority and think scientifically are given by exposing the reader to research methodology, mathematics, history, integrated science content, educational research, and philosophy as roots to science literacy. The excitement and optimism science promises for the future is juxtaposed with threats to scientific integrity and declining results of national science efforts.

*Science & Society* may be used as a core text or supplementary text in courses for any introductory field of science: biology, chemistry, geology, or physics. It presents an integrated approach to how each field of science is developed and fits within larger areas of study. Its format is intended to treat the nature of science in general and the major principles within each discipline. The book is designed to help readers develop their own thinking about science. It is able to do this because *Science & Society* is one of the few texts that address all of the contributing disciplines that create science as a rich subject. Its history, philosophy, mathematics, sociology, content, and science education underpinnings present the reader with a view of science as multifaceted and yet simply interdisciplinary.

It provides an overview of science without anchors to any one discipline. It discusses the scientific method in terms of research methodology but also the mathematics that gives science power to effect change. Both the philosophy and history of science sections embed the readers' perspective with a reflection on scientific roots. The influence of science on societal development is a main focus of the text and presents the sympatric relationships of science areas within a sociological framework. Science is portrayed as its own community, with rewards and integrity issues, hindrances, and a competing array of pseudosciences. Practical advice and guidelines for becoming a scientist, pursuing a career in science, and teaching science is also given. Original research and contemporary works on improving science reasoning and retention in science majors is given in several sections of the text.

Strategies to develop critical thinking skills are presented and applied to the content material.

One part of science literacy comprises facts and principles in science. This text also presents a content knowledge base for the reader. An overview of the salient aspects of science content in the four branches is also presented. The text as a whole provides a foundation of knowledge in science to successfully begin study for a scientific career. The reader is exposed to several content applications throughout the text. *Science & Society* is meant to complement more comprehensive, fact-driven texts by presenting science as it progresses within our civilization.

The book is an excellent resource both for students entering a science career and pre-service teachers who need scientific thinking to guide their own classrooms. Secondary and elementary education science methods courses need a book that brings out science process embedded in content and methods. *Science & Society* does this by providing teaching methods with the philosophical, historical, mathematical, and sociological foundations necessary to foster effective science instruction. The viewpoints, content, and applications of this text conform to the standards-based reform efforts to improve science teaching.

The use of this book in the often idiosyncratic interdisciplinary science courses allows liberal use of instructor materials to augment the text and personalize the course for each instructor. This text is provocative and argumentation-based to stimulate good, meaningful in-class discussion. It is broad enough to allow for flexibility in course designs but sets forth a valuable core curriculum. The extensive references at the end of each chapter in the text enable the instructor to research and integrate course materials with their own research articles.

Science is shown, throughout the book, as a dynamic part of a changing society. In short, the interdisciplinary approach to learning science leads the reader to real truths behind many natural phenomena. It develops an appreciation for the way in which we gain scientific knowledge. The purpose of the book is thus to excite the reader's innate interests in the scientific process and to recruit good people into science. Our society needs this book.

## Acknowledgements

I thank my wife and editor, Amy E. Daempfle, Ph.D., who is my inspiration. I thank my father, Tobias Daempfle, who has had great impact on my life and whose conversations have contributed significantly to this book.

I also thank my teachers of the past, who helped form the foundations of thought to create this book. Their ideas were interdisciplinary, creative, and forward thinking. Their teaching was profound. They gave a great deal, more than they can know: Carole Demian, Robert F. Pospisil, John P. Rosson, Wendell W. Frye, William M. Elliott,

Audrey B. Champagne, Robert F. McMorris, and Margaret Kirwin. Let their teaching touch eternity. To my students, whom I have had the honor of teaching and sharing ideas. Let our paths cross again through this work. Thanks also to the supportive structure of the SUNY system and SUNY Delhi.

I appreciate the thoughtful reviewers of the early drafts of the text manuscript who helped make this book better:

Chris De Pree, Agnes Scott College
Andy Digh, Ph.D, Mercer University
Dr. Paul Gier, Huntingdon College
Olivia Harriott, Ph.D, Fairfield University
Marlene M. Hurley, Empire State College

Special thanks to the exceptional staff at Jones & Bartlett Learning, including Megan Turner, Alex Schab, Erin O'Connor, Rachel Isaacs, and Ashley Dos Santos for their patience and assistance in helping me through the process of developing this text.

## About the Author

Peter A. Daempfle, Ph.D. has taught biology, anatomy and physiology, human genetics, science issues, and science education courses for 20 years. Dr. Daempfle has held faculty positions at Hobart and William Smith Colleges, Western New England University, and is currently an associate professor of biology in the State University of New York, College of Technology at Delhi. He earned his Ph.D. in Science Education and his M.S. in Biology at the University at Albany, State University of New York; M.S. in Education from the College of Saint Rose; and B.A. in Biology from Hartwick College. He was class valedictorian of both Forest Hills High School, Queens, NY in 1988 and Hartwick College, Oneonta, NY in 1992 and graduated *summa cum laude* with departmental distinction in biology and German. Born in the Ridgewood section of Brooklyn, NY, in 1970, Dr. Daempfle was a child of German immigrants.

Dr. Daempfle was the first science education researcher to use qualitative and quantitative approaches to study the academic transition between secondary and post-secondary biology programs. His journal articles are cited extensively and used in contemporary studies throughout the science education literature. A scholar in his field, Dr. Daempfle has authored several journal articles, various science reviews, a laboratory manual, *Introduction to Anatomy and Physiology,* and lectures to scientific and general audiences.

From 2001–2009 he was a science advisor to the Bush Administration's No Child Left Behind Act (NCLB). He focused on science content applications to psychometrics and test design in relation to standards development. He is known in science literature for publications focusing on the development of scientific reasoning, retention of students in science, studying the tenuous transition between secondary and post-secondary science programs, and human biology and microbiological applications. This new text contributes to the effort begun by the NCLB Act to improve national science literacy and advance the importance of scientific thinking.

# CHAPTER 1

# Introduction

## A Story of Science

While sitting on the beach at the Pomeranian seacoast, a Capuchin monk once saw a small girl working very hard using a soup ladle to take water out of the ocean. After watching the girl work a long time, the monk went up to the girl and asked her, "What are you doing, working so hard?" The girl replied, "I am trying to empty the ocean so I can see what is underneath." So too is human knowledge, thought the monk. "Just as the girl will never empty the ocean, humans will never understand all there is to know. We are merely a grain of sand on the beach in the largeness of the universe. Many have been driven mad trying to figure it all out." Despite thinking this, the monk became very hopeful. He then went over to the girl to help her empty the ocean.[1]

All of us have felt the excitement of the monk and child in the story. Perhaps in wondering what (or who) is on the stars or how a computer works we elevate our thoughts. Our determination to discover the unknown and explore the truth about life makes each of us a scientist. As in the story, science begins by trying to unravel

1

the vastness of information in the universe. There are obvious limits on the capacity of science to discover the unknown. There is also a human desire to tame it, as seen in the story of the girl and then the monk. Should we try? We are unable not to. This is where science begins.

Unfortunately in our society, the word *science* often evokes a variety of possible connotations: lists of long and hard words, asocial people in lab coats, or maybe even an *experiment* or two that was done in your middle school classroom. Traditionally, we are exposed to science in our schooling. None of this truly captures the essence of science.

A definition of science is difficult to develop. Even scientists have trouble agreeing on a definition because there are so many facets to the word "science." It is a way of knowing—a way of thinking about the world. It is a process by which we understand the universe. It is, in its purest form, a way of arriving at some truth. But knowing, understanding, and truth are all hard to define.

## What Is Science?

British philosopher William Whewell (1794–1866) first popularized the word *science* in 1851. His simple definition stated that science is a study of the natural world. The word *science* derives from the Latin word "scientia," meaning "knowledge" or "knowing."[2] As such, through much of modern human history, people who studied the natural world were known as natural philosophers. Aristotle, Plato, Socrates, Copernicus, and Newton were examples of scientists long before the term science was coined. Indeed, their contributions as natural philosophers to the scientific world were vitally important.

A modern definition of science consists of three parts. First, it is a *body of knowledge* about the natural world. There are known facts and principles on which knowing about the universe is based. Second, science is a *method*, or a way of finding out about some truth. The scientific method has required elements that make it different from other, nonscience explorations. Third, science requires *reasoning* to understand its base knowledge and apply new discoveries derived from using the scientific method. Scientific reasoning was once called "commonsense." It is comprised of rationalism (a link between cause and effect), openness to investigation and inexactness, and an ability to recognize patterns and solve problems. Reasoning is important in everyday life and is the foundation for scientific advancement. These three essentials are the foundation on which this book is organized.

The text is an introduction to how scientists think. It illustrates the role of science in leading human progress and cautions of the dangers involved, which could lead to our destruction. It discusses scientific thought and innovation with consideration for historical and philosophical contexts and their influences. Finally, it encourages the reader to develop his or her own philosophy about science, the world, and even our existence.

# The Nature of Science

Science is often taught as separate, discreet entities. Traditionally, science is divided into the four general areas: *physics* (study of interactions of matter and energy), *chemistry* (study of composition and reactivity of substances), *biology* (study of life), and *geology* (study of earth, space, and their processes). It is true that one builds upon the other, in the order given. However, recognizing that all sciences are *interrelated* gives us a fuller understanding of phenomena.

Understanding any natural phenomenon requires knowledge in all areas of the sciences. To illustrate: Physics deals with natural laws such as gravity and intermolecular forces of motion. In medicine, treatments for high blood pressure are derived from an understanding of these laws. Laws of physics dictate that all matter, even solids, is in a constant state of motion due to an intrinsic *kinetic* or moving energy. When objects are in motion close together they bounce off each other. Kinetic energy keeps objects moving and hitting one another so that they spread out ever farther and farther apart. Chemists call this process *diffusion*, the movement of particles from a higher concentration to a lower concentration until an equilibrium (even spread) is reached. When water moves along a membrane in this fashion, it is termed *osmosis*.

To bring blood pressure back into the picture, most of us know that a high salt diet contributes to high blood pressure. Many people, however, are unfamiliar with the science that tells us why. The answer is based on the above discussion of physical and chemical principles. The movement of salts and water can be traced in **Figure 1.1**.

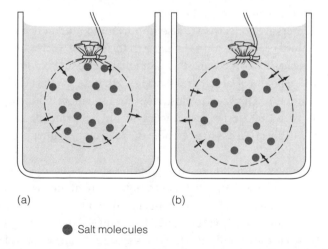

(a)                              (b)

● Salt molecules

**Figure 1.1**   Osmosis: Osmosis is the movement of water molecules from a higher concentration to a lower concentration across a membrane. (a) A bag of saltwater is immersed into a solution of pure water. (b) Afterwards, water diffuses into the bag, causing it to expand and increasing pressure within the bag.

If a diet is high in salt, more salt will make its way into a person's bloodstream. As Figure 1.1 shows, water follows salt molecules into the bag, causing it to expand. As a result, there is more pressure on the walls of the bag. Similarly, water following salt into blood vessels increases pressure on the cells lining blood vessel walls. Diffusion will move water from high concentration areas (in body cells) to where it is lower (in the vessels) in a person with a high salt diet. When water enters the vessels, the higher blood pressure may cause damage. This is responsible for many health ailments including *arteriosclerosis* (hardening of the arteries) and related heart attacks and strokes.

Biologists, applying nonbiology areas, research different chemicals to treat high blood pressure. The human body, a complex living structure, responds to drugs for high blood pressure treatment in a few ways. Some drugs (*vasodilators*) expand the vessels to lessen pressure and some (*diuretics*) reduce the amount of water in the vessels by increasing urination. While the human body does respond to the chemical and physical laws of science in the treatment of blood pressure, sometimes patients actually have an opposite effect or no effect from the medication. This demonstrates the possibility of error inherent in the process of science. There is always a percentage chance that things may not work out as planned in a scientific investigation or intervention. Frequently evident in medical research, these nonconformities can be particularly frustrating. Mathematicians study the probabilities of scientific error and analyze the results in order to further scientific research. The human body is more complex than we can imagine and there is always more to learn. This is cold comfort for the patient, whose body may not be responding as expected to the known laws of physics, chemistry, biology, and mathematics.

High blood pressure concerns are also important to human health and society, with nearly one-third of all U.S. adults living with the disease. Increasing obesity rates in Western society has led to, among other ailments, increasing blood pressure in the populace. After all, a pound of fat requires the body to build 1,000 feet (304.8 meters) of blood vessels to supply it. This leads to more resistance and more blood pressure needed to overcome the extra 1,000 feet of resistance. This physics principle has major effects on heart health as the heart needs to overcome the load of the higher pressure of blood exerted on the heart.

Ultimately, living systems attempt to maintain a steady state or constancy around a normal set of conditions. This struggle to maintain balance within the body is termed *homeostasis*, and an imbalance of this is when disease begins. In the case of blood pressure, the human body is best suited for a range of pressure hovering around 120 to 80 mmHg when measured with a blood pressure cuff. The vessels of the body are structured to withstand this range of pressure over the lifetime of a human. However, failure of the structures due to higher pressures results in conditions such as *aneurisms* (ballooning out and breaking of vessels), coronary artery disease (damage to the lining of the heart), which may result in stroke or heart attack, or weakening hearts (due to the high pressure wearing out parts of the heart such as valves).

Where do the geosciences enter this example? The Earth also maintains a kind of homeostasis through regulation of its water and salt balances. Humans are creatures inherently linked to Earth water systems. We are almost 80% water by composition, and require continual inputs and outputs of fluids and salts. Have you ever wondered why every morning you are thirsty and yet are compelled to urinate? We are intrinsically linked to the sea. Principles of *osmoregulation* (water pressure control) require that we have the salt of the sea to regulate our internal water balance. Our human bodies require the right balance of salt in order to live and function most efficiently. Earth provides sources of salt that are available to us for consumption so that we can achieve this balance. Similarly, Earth also requires the right balance of salt in order to regulate ideal planetary conditions. For example, variations in ocean saltwater concentration and temperature are responsible for creating deep water ocean currents that in turn help regulate global surface temperatures. Pockets of sinking ocean surface water that fall to the ocean floor drive this process. This salt-based self-regulation of the environment contributes to the ideal temperature conditions for us to exist as humans on the Earth, as shown in **Figure 1.2**.

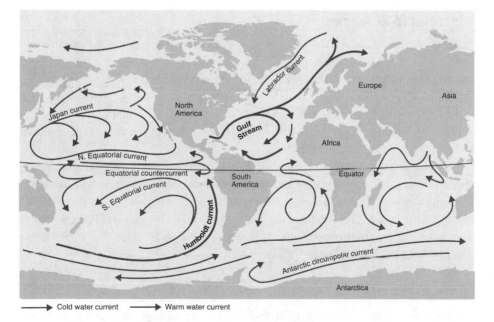

**Figure 1.2**    Major Ocean Currents

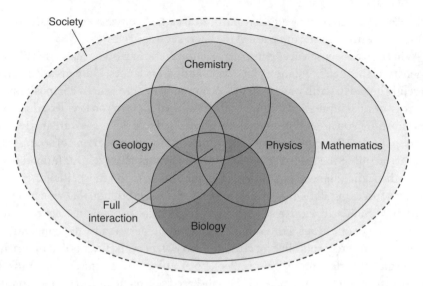

**Figure 1.3**    Interactions within the Scientific Paradigm

## Interrelated Sciences

As demonstrated by the blood pressure example, understanding the "why" questions of how science works is complex. It requires an interaction of not only the four sciences but must be understood within the context of mathematics and society as a whole. Blood pressure research depends on the mathematical analyses of the results to help society use the right treatments for this pervasive problem. The illustration below in **Figure 1.3** pictorially represents these relationships between various domains in understanding scientific phenomena.

Depicted in the model above are the different branches of knowledge overlapping and interacting to form one way of knowing. This blur of the distinction between areas of thought was well discussed by the geographer Halford Mackinder (1861–1947) in 1887 who stated, "The truth of the matter is that the bounds of all the sciences must naturally be compromises; knowledge . . . is one. Its division into subjects is a concession to human weakness."[3]

## Characteristics of Science

While science is often seen as divided, its branches certainly share common features. There are five general characteristics of science. First, science is always *empirical*. That is, it is based on observations and first-hand experience. Empirical data is knowledge gained through the five senses: seeing, hearing, tasting, touching, and smelling. Science is based on what is observable in the universe. Observations are measurable and measured data is difficult to dispute.

Of course, the observable must be caused by natural rather than supernatural phenomena. A second shared characteristic is the assertion that science exists only in the *natural world*, in which conclusions are based on acceptable, logical principles. The biologist does not view a disease, such as liver cancer, as a punishment from God and astronomers do not see Heaven between the moon and Earth. Various forms of "pseudoscience," such as intelligent design, UFO conspiracy theory, and astrology, do not conform to natural explanatory mechanisms and rely on supernatural belief. Pseudoscience as a social replacement and/or alternative to science will be addressed later in this text. Pseudosciences are based on belief; belief is a difficult concept to demarcate within the scientific community. It takes a certain amount of belief to develop ideas in science but belief must be backed up with empirical data obtained from the natural world. **Figure 1.4** represents the related relationship between belief and science. There has always been a fine line between the two. One cannot exist without the other but the goal of science is to perpetually expand into the field of belief. As seen in the story at the start of the chapter, the goal of science is not to fight belief but to gather evidence to support or refute beliefs based on reason and investigation.

A third characteristic of science is its need for *repeatability*. Scientists should always be able to, under the same conditions as the original study, repeat an investigation and come up with the same conclusions. Is this so in your experience with scientific endeavors? "At times" may be your answer. Of course, you have seen different diets and their studies come and go, treatments for diseases change, and claims about global warming vary. This is because science changes. Even though experiments should

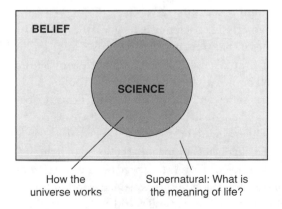

**Figure 1.4**    Relationship Between Belief and Science

have repeatability, changes in the experiment can and often should lead to different results. Science, however, is not anecdotal. That is, it should not rely on individual observations, scattered like UFO sightings. Instead, it should be based on repeatable patterns. Repeatability with consistent results is what strengthens reliability in science.

Thus, the fourth characteristic of science is that it is *testable* via experimentation. An experiment is really a planned intervention. It is a study that analyzes the effects of a particular variable. Let's review an experiment I conducted in embryology class in college. The effects of retinoic acid (a skin cream component) on frog embryo development were observed. A control group, which allowed the embryos to develop under normal conditions, was established. Next, an experimental group of embryos that developed in varying concentrations of retinoic acid was created. The results revealed deformities in the experimental group. One embryo had its axis reversed, meaning that its head was in its butt region and its butt was in the head region. The control embryos all appeared normal.

The *independent variable* is the condition that the experimenter alters. In the embryology experiment, what did I change? Yes, the retinoic acid levels. The *dependent variable* is that which is modified as a result of the independent variable having been changed. In other words, it varies or depends on the independent variable. What was the dependent variable for the embryology experiment? Yes, the anatomical changes in the embryo. Simply put, the dependent variable reveals the results of the experiment. Because it isolates the effects of one particular condition, a carefully designed experiment gives the best empirical data of any type of investigation. An experiment seeks to keep all of the variables affecting the results the same except for the independent variable. These are termed *control variables*, which are those factors that remain the same among all of the groups under study. The strength of a study is in part based on how controlled the experiment is; the better the controls, the stronger the study results. We will explore research methodology later in the text to help in developing and critiquing scientific research.

The fifth characteristic of science is its goal of establishing *generalities*, that is, finding principles or natural laws that are universally applicable. Such findings should apply to all circumstances. For example, *Newton's laws of motion* apply to the movement of all bodies in the universe, not just some. As discussed previously, diffusion of particles relies on Newton's laws and forms the basis of application in medicine, geology, and chemistry. Of course, there are exceptions to laws. Einstein's theory of relativity incorporated the effects of speed and time for an object. Simply stated, as the speed of an object approaches the speed of light, it slows its progression in time. Newton's laws do not address the speed of light and time as variables. Newton's physics stops working under conditions explained by *relativity*. However, for the most part, Newton's laws apply in our natural world and generally work. Science involves general principles (albeit with exceptions) to rely upon in understanding phenomena.

## Relations Between Science and Society

Science affects society and society affects science. Examples illustrating how science influences social development are too numerous to list comprehensively. The discovery of radioactivity, inventions, and developments in medicine, evolutionary theory, and global climate change data are just a few scientific leaps that have led to major shifts in thinking in society. Let's explore how these developments have affected your own thinking.

### Thought Questions

If I were to ask you where you stand on the issues below, which side would you choose and why?

1. Atomic bomb: Good or bad discovery?
2. Ethics in medicine: Would you use prenatal testing to support a decision to abort a fetus that tests positive for Down syndrome?
3. Evolutionary theory: Is this an acceptable explanation for how humans came into existence or not? and
4. Global warming: Is it due to human activity or not?

Obviously, there are many facets to these questions. We all probably fall into some range between yes and no in answering these questions. All have had profound impacts on our society.

### *Possible reflections on the questions:*

1. Can you think of ways radioactivity has affected our lives? Yes, there is great fear that a nuclear attack could destroy humanity. Having no fear of this is foolhardy. But the benefits of nuclear medicine on cancer treatments, political effects of preventing conventional wars, and decontaminating water sources can all be attributed to this discovery.
2. The ability to test human fetuses has led to a host of bioethical issues in medicine. If there is a choice to terminate a pregnancy, then where is the line drawn for basing it on prenatal tests? At what point, if any, does the individual make a decision? Spina bifida, Down syndrome, Tay Sach's disease, or, with potential genetic testing, Huntington's disease or even Alzheimer's. Is that choice acceptable under any circumstances?
3. Evolutionary theory has been one of the most contested postulates in recent history. There is a preponderance of evidence supporting evolution but it is a hot-button issue in society because many feel that it questions God's role in creating life. If we simply evolved from chemicals, so the reasoning follows, then the question becomes: "Is there a need for a creator God?" A popular competing view by intelligent designers asks: If evolution took place

how did it become so "irreducibly complex" (which is the idea that life is too complex to have simply formed spontaneously)? Questioning evolution is popular because it appeals to a public with a nonscientific background. Alternatively, the great biologist Theodosius Dobzhansky (1900–1975) argues that "Nothing makes sense in biology except in the light of evolution."[4] Given enough time, evolutionists argue that mutations led to certain benefits for some species to help them survive better and thus, "evolve." There is abundant evidence for evolution. Our minds must be kept open because scientific views change based on new information. Who do we believe?

4. Finally, global climate changes have been demonstrated since the start of the Industrial Revolution. There is strong evidence that human activity is increasing atmospheric $CO_2$ levels and global temperature changes are very much in tandem with fluctuations in $CO_2$, as shown in **Figure 1.5**. While most scientists support this perspective, there is always the possibility that new data may emerge to undermine this viewpoint and it remains a studied phenomenon. Global climate change is debated in the public domain and is an area of sociopolitical controversy. Often cited in public debate, there have been natural fluctuations in temperature throughout geologic time, long before human inputs. The Earth's temperature cycles every 100,000 years, as shown in **Figure 1.6**. Ice core samples taken by geologists indicate even greater natural historical variations in temperatures than at present. After all, the *Little Ice Age* in the 1700s was related not to human activity, but to volcanic ash collecting in the atmosphere and decreasing direct sunlight. This does not mean that humans are not responsible for today's climate changes. The public's main question is: Are humans responsible for the current increases?

As indicated in the above examples, science has affected society and society has affected scientific thinking. Scientific development does not occur in a social vacuum. Scientists themselves are influenced by the views and motivations held by the societies in which they live. Money, for example, often drives progress in science. Consider at what times historically momentous scientific advances occurred? Wartime is certainly one of these circumstances. Because money is diverted to developing technology to fight the enemy, rapid progression in science areas is frequently a fortunate product of war.

To illustrate, sonar was first developed during World War II to find German submarines. Sonar uses high frequency sound waves to detect objects beneath the sea. Medical advances and applications using the same high frequency sound waves soon followed. Echocardiograms to detect heart activity and ultrasound to detect internal structures such as fetal anatomy are examples of sonar-based technology used in medicine today. Society's effect on science is a major theme threaded throughout this text.

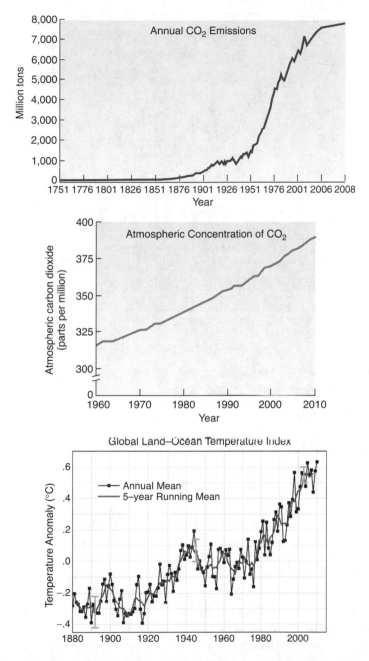

**Figure 1.5** Global Carbon Dioxide Emissions and Global Temperature. (a) Annual carbon emissions from fossil fuel burning, 1751–2004. (Data from UN, BP, DOE, and IEA) (b) Annual mean carbon dioxide levels in the atmosphere have risen dramatically since 1960. (Data from Scripps Institute of Oceanography.) (c) Graph of average global temperature since 1950. (Data from GISS, BP, IEA, CDIAC, DOE, and Scripps Institute of Oceanography.)

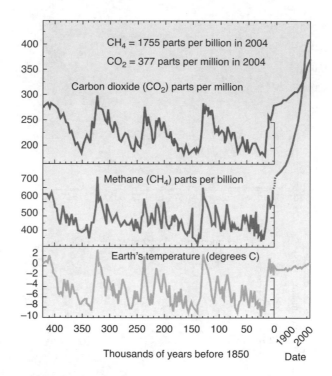

**Figure 1.6** Natural Cycles. Methane and Carbon Dioxide gases undergo natural cycling in tandem with global temperatures

Reproduced with kind permission from Springer Science+Business Media: Clim. Change, A slippery slope: How much global warming constitutes "dangerous anthropogenic interference?" vol. 68, 2005, pp. 269–279, J. E. Hansen.

## Science Changes

As shown in the above description of the development of sonar, science changes based on the need for new applications. High frequency sound wave applications transformed and progressed during peacetime use from detecting and destroying submarines to saving the lives of cancer and heart patients. Science also changes with the revelation of new facts and data. This can often be frustrating for the public. I remember as a child, my aunt Pauline had an old exercise machine that jiggled you as you stood there. You did no exercise but it was supposed to make you lose weight. Obviously, exercise physiology has progressed since that era of jiggling. It is scientifically shown that nutrition and healthy eating are more important for human health than simple jiggling. That being said, there are a host of dangerous gimmicks on sale to help weight loss, from stretch machines to diet pills that have been linked to heart valve damage. While science has changed, a frustrated public needs to have the tools necessary to decipher between valid and invalid science claims. The upcoming chapters in this text give those skills to evaluate the many changes claimed by research studies in the media.

# Spontaneous Generation

Science is a complex interaction of changes in data received, new scientific outlooks and changes in society. For example, consider the historical development of thought on *spontaneous generation*, which is the assertion that life can emerge from non-living substances. In 1785, the Italian monk and scientist Lazaro Spallanzani (1729–1799) studied the role of semen in fertilization. He disproved a "vapor" hypothesis (a proposed explanation for a phenomenon) that claimed semen could fertilize an egg without penetration. We now know this to be wrong. Imagine if sitting too close to someone resulted in a pregnancy! Spallanzani studied toad semen with a variety of experiments to show that semen vapor could not simply move through the air and fertilize an egg. In one experiment, he showed that semen lost its reproductive power when traveling through a cotton ball. In another, the semen could not fertilize eggs through a paper filter. However, in a wet cotton ball the semen was able to fertilize with full force. He concluded that semen was the agent of fertilization.

However, Spallanzani held firmly onto his belief that it was semen and not sperm that acted as the agent. Despite experimental evidence to the contrary, Spallanzani had already decided that semen itself fertilized eggs. It was not for another century that sperm was determined to be the reproductive agent. While this evidence now obviously shows sperm's significance, Spallanzani rigidly overlooked his own data. Thus, scientists' *pet hypotheses*, or views held prior to an experiment, can contaminate the interpretation of data.

Spallanzani based his research on another scientist, Francesco Redi (1636–1697), who looked at how new life arises. Redi, an Italian naturalist, was the first to disprove *spontaneous generation*. It is generally accepted that food left out in the open soon contains maggots (the larval stages of flies). Seventeenth century scientists hypothesized that the organic matter in food automatically generated maggots and all associated life when coming into contact with air. Redi devised an experiment in which a piece of meat was placed in a glass jar and the jar was covered with gauze. This allowed air flow but no other agent to get to the meat. The control jar was uncovered, of course. At the conclusion of the experiment the gauze-covered jar did not have maggots but the uncovered jar did. Redi concluded that there was some other agent causing the maggots and not the meat itself as the source of life. We now know that flies are that agent but in the seventeenth century there were no microscopes to view fly eggs. This was a first step in disproving the idea of spontaneous generation.

Nonetheless, biologists in the 1860s were again debating the issue in terms of milk and beer souring due to bacteria. French biologist Felix Pouchet (1800–1873) argued that although bacteria did reproduce themselves, they initially formed due to the right combination of organic materials. He believed in spontaneous generation. Pouchet performed experiments that heated flasks of hay infusions to 100°C and then sealed them. He tried to show that bacteria would grow even though sterilization destroyed all of the organisms present. Despite this "sterilization," bacteria formed in the flasks and Pouchet concluded that these organisms arose from a good mixture of materials in the hay infusions. He tried many times to sterilize the environment in the flasks but within a short period of time, he observed a sea of bacteria in the sampled hay infusions under the microscope. He thus concluded that life could arise from nonlife without a parent organism.

Louis Pasteur (1833–1895) proposed an alternate hypothesis, arguing that bacteria were everywhere and therefore contaminating Pouchet's experiment when he placed the lids onto the jars. Pasteur boiled beef broth in specially designed long-necked flasks to keep the broth within free from bacterial growth. Air was able to get through to the broth but the lower part of the neck served to trap the heavier dust particles and microbes. His flask design is shown in **Figure 1.7**.

Without an external agent, Pasteur reasoned, spontaneous generation would not occur. He was correct; the "trap" in the neck kept out microbes. Then, when Pasteur tipped the flask to allow the broth to touch the trap, bacteria appeared in the broth in a few days, as his hypothesis predicted. This rejection of Pouchet's viewpoints led to Pasteur's membership and award from the French Academy of Science.[2] His experiment showed how ingenuity and reasoning led to a final disproof of spontaneous generation and the birth of microbiology as a discipline.

Sociologically, Pasteur was a devout Roman Catholic. He performed the experiment to validate the religious premises emphasizing the sanctity of life. If life could come from nonlife, the significance of God in creating life was in question. His disproof of spontaneous generation was a movement against atheism and toward a resurgence of religious importance in the 1800s. Pasteur himself promoted his work as an example of pure science, unbiased and uninfluenced by society. His detachment is still in question.

The Pasteur-Pouchet case illustrates how scientific arguments continue through the centuries. Pasteur's work led to the *germ theory* of biology (which advocates sterile techniques to prevent microbial disease spread) and thus vast improvements in medicine due to widespread use of sterile technique. However, some of his results were questioned. In 1953, physical chemists Stanley Miller (1900–2007) and Harold Urey (1893–1981) devised an experiment demonstrating that life could have formed from the right mixture of chemicals. Conditions used were hypothesized to be the same as found in early Earth (e.g., methane, hydrogen sulfide, hydrogen gas, and water vapor). The experiment in **Figure 1.8** shows the set-up of Miller and Urey's investigation.

A set of organic molecules, including sugars, fats, proteins, and genetic material, was produced from the simulation, which could give rise to life. Pouchet's hypothesis re-emerged successfully. The debate on the origin of our existence continues.

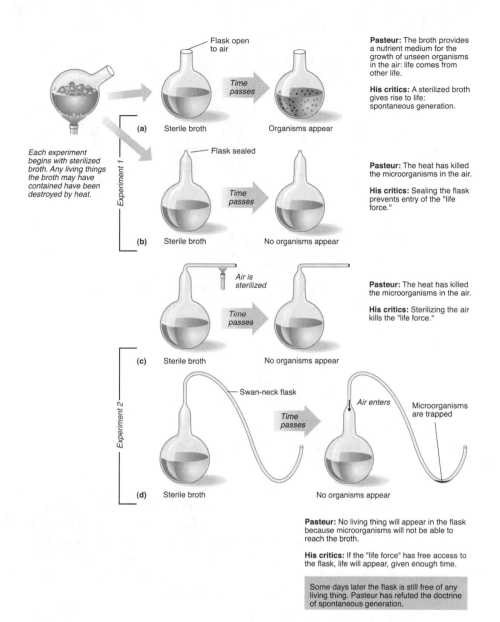

Flask open to air

*Time passes*

**Pasteur:** The broth provides a nutrient medium for the growth of unseen organisms in the air: life comes from other life.

**His critics:** A sterilized broth gives rise to life: spontaneous generation.

**(a)**    Sterile broth    Organisms appear

*Each experiment begins with sterilized broth. Any living things the broth may have contained have been destroyed by heat.*

Flask sealed

*Time passes*

**Pasteur:** The heat has killed the microorganisms in the air.

**His critics:** Sealing the flask prevents entry of the "life force."

**(b)**    Sterile broth    No organisms appear

Air is sterilized

*Time passes*

**Pasteur:** The heat has killed the microorganisms in the air.

**His critics:** Sterilizing the air kills the "life force."

**(c)**    Sterile broth    No organisms appear

Swan-neck flask

*Time passes*

*Air enters*    Microorganisms are trapped

**(d)**    Sterile broth    No organisms appear

**Pasteur:** No living thing will appear in the flask because microorganisms will not be able to reach the broth.

**His critics:** If the "life force" has free access to the flask, life will appear, given enough time.

Some days later the flask is still free of any living thing. Pasteur has refuted the doctrine of spontaneous generation.

Experiment 1

Experiment 2

**Figure 1.7**    Pasteur's Experiment

It is clear from these historic examples that science is undoubtedly influenced by society and does not occur in a social vacuum. Scientists are human beings and bring bias to their investigations, for better and for worse. Can you give examples of bias that occurs in science in our modern society?

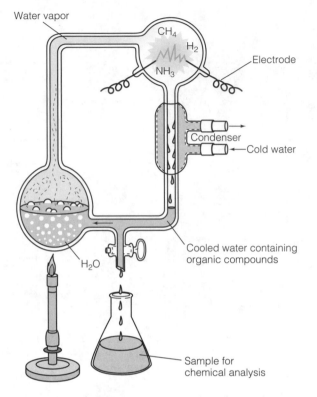

**Figure 1.8** Miller and Urey Apparatus. This device showed tat organic molecules could be produced from the chemical components of Earth's early atmosphere

## Science as Inquiry

Science is a way of thinking; a way of knowing. *Inquiry* is defined as the logic behind science and the way of thinking to discover its phenomena. Inquiry has all the characteristics of science; it follows a logical sequence of attempting to get at truth but also is more haphazard, backtracking in ideas and reformulating strategies. Science based on inquiry is very much like the game of chess. Thomas Henry Huxley (1825–1895) described it as such in the following excerpt from 1868:

> The chessboard is the world, the pieces are the phenomena of the universe, and the rules of the game are what we call the laws of Nature. The player on the other side is hidden from us. We know that his play is always fair, just and patient. But we also know, to our cost, that he never overlooks a mistake; or makes the smallest allowance for ignorance. To the man who plays well, the highest stakes are paid, with that sort of overflowing generosity with which the strong shows delight in strength. And one who plays ill is checkmated— without haste, but without remorse.[6]

While other chapters discuss the philosophy of science and its methodology, the essence of science is well captured in this analogy to a chess game. Science can be philosophized about and detailed but the soundness of an inquiry is akin to a chess player's skill. It is complex and multifaceted. Chess and inquiry both advance in an uncertain manner, with creativity and expertise that are hard to define.

Science inquiry, like chess, requires both *induction* and *deduction*. Induction is much like how the character-detective Sherlock Holmes investigates a scene. He looks at all of the facts and pieces them together to form some sort of conclusion or pattern statement. It is really a gift to be adept at induction. The relationship between induction and deduction is given in **Figure 1.9**.

Induction requires creativity and a way of thinking that looks at all of the elements in a situation. In the example of Pasteur, he came up with a view of the problem and solutions from the data he observed. The Germans call this "world view" or *weltanschauung*. Americans call it a *paradigm*, or a way or model for looking at things. Pasteur used inductive reasoning from other scientists' data to develop his weltanschauung and proceeded toward the germ theory of biology (pattern statements).

Deductive reasoning is the movement from the pattern statement by testing it and coming up with facts or data as results, as shown in Figure 1.9. Pasteur's results showing microbes under certain conditions led to further inductions and deductions by experimenters coming after him. They used his facts to confirm the existence of microbes, spores, and viruses as agents of life.

Consider an activity to compare induction and deduction. If a group of people is asked to find a million of something inside a given building, they are engaging in deduction. They are given a statement and asked to support it or fail to do so. If another group is told to form a statement about a million of something in a building, it is carrying out an induction because the group is asked to gather data and form a conclusion. The induction is more open-ended and requires, in some ways, more creativity. The

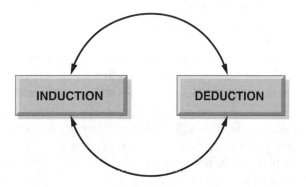

**Figure 1.9**   Induction and Deduction

deduction is more proscribed whereby the group must either support the statement or fail in doing so. It still requires creativity in gathering data but is less free in its possible responses. One group of artists once demonstrated that there was a million of something by showing a blade of grass as a number one and adding six zeros represented as pebbles found in the basement. Together it created a $1000000 = 1$ million. Individual creativity and flexibility is an important part of inquiry and the scientific method.

The movement between induction and deduction is the essence of science inquiry. Uncertainty is inherent in this process. With every deduction there is another induction that can disprove the first. Karl Popper (1902–1994), a famous philosopher on science as a process, best expressed this stating, "Science advances only through disproof of alternate, working hypotheses." These hypotheses are the "guesses" scientists make during their inquiries. They are often changing and never certain. Hypotheses are never proven, only falsified. One either supports or fails to support a hypothesis. No matter how many times a hypothesis is shown to be true, scientists never know if there will be evidence to the contrary to come forward. Consider attempting to prove that "all swans are white." Even if a million swans are white, it does not prove the hypothesis that all swans are white. A nonwhite swan may exist that has not yet been found.[7]

Medical diagnoses are often developed through this process of inquiry. Initial diagnoses are frequently refuted. Consider a cough. It is likely a cold so the doctor prescribes rest and fluids or an antibiotic. If that fails, then a process of continued tests ensues that could eventually point to lung cancer. Rarely is cancer a first diagnosis. But, a doctor should never say, "You are definitely OK" either. While frustrating for a patient to hear, this is science and there could always be a test or evidence to show the contrary.

Consider the statement "Cigarettes do NOT cause lung cancer." This is true. One might probably be shocked by that assertion. What if it were to further be explained that in studies of 100,000 smokers versus nonsmokers, no statistical difference in lung cancer rates were shown, given a t-test (statistical test to show significance)? In addition, what percentage of smokers actually gets lung cancer? One may guess 10 or 20%. No. Actually only 1% or 1 in 100 smokers develop lung cancer. So light up? These are valid statistics used by tobacco companies for decades. This is why one can never say cigarettes are "proven" to "cause" lung cancer.

Now, contemplate the alternate data. What percentage of people with lung cancer were smokers? Yes, quite high (90%)—and statistically significant. When the numbers are presented in this manner, the link between smoking and lung cancer is more clearly demonstrated. Obviously, research and mathematics can be deceptive and are used to manipulate decision-making. This will be discussed in more detail in other chapters. Both sets of data are true but each present a different picture.

The public could accept the first set of data without knowing about the second set. This could give a false sense of security about the negative impacts of smoking and mask information about the many health-related problems associated with smoking: heart disease, throat cancer, emphysema, asthma, wrinkles, and oral cancers. However, while the second set of data show clearly that there is a relationship between smoking and lung cancer, science is always uncertain and the relationship is never *proven*. Scientific data can be misleading and need to be looked at closely to reach further to the truth of a matter.

## Dichotomies

All natural phenomena have opposing sides: light vs. dark, positive vs. negative, proton vs. electron, sun vs. moon, positive force vs. negative force, good vs. evil. Much like nature, scientific study involves opposing sets of viewpoints. This text is centered on analyzing scientific arguments that have opposing sides to an issue or hypothesis in question. Scientific thinking involves analyzing facts (reasoning and argumentation) and forming conclusions to study these kinds of contrasts. Science and its phenomena are filled with a strange series of tensions that oppose each other, called *dichotomies*.

In fact, consider the moon and the sun as examples. The moon and the sun appear the same size as one another upon simple visual observation from Earth. Yet we know that the sun is much larger. The diameter of the sun is 867,000 miles (1,395,301 km) and the diameter of the moon is 2,155 miles (3,468 km). The sun is actually about 400 times larger than the moon in diameter. Then why do they appear the same size? The sun is also about 400 times farther away than the moon is from Earth. The moon is on average 238,857 miles (384,403 km) from the Earth while the sun is a whopping 93,000,000 miles (149,668,992 km) away. Differences in the distances and diameters are opposite but equal. Thus, the physical properties of the moon and sun balance to make the two appear equal in size. This is a freak example but shows how observations can sometimes be misleading. In fact, a skeptical outlook, based on measurements and methods, can combat false scientific conclusions. An eclipse in **Figure 1.10** shows the relative size of the moon and sun when they are superimposed upon each other during a solar eclipse.

Courtesy of Alex York

**Figure 1.10**   Solar Eclipse: During a total solar eclipse, the light of the sun's outer atmosphere—the corona—is visible behind the moon. This photo of the June 30, 1992 eclipse was taken from the window of a DC-10 30,000 feet above the ground.

In other chapters, there is an emphasis shown on the juxtaposition of different scientists' ideas with each other. This conflict within the scientific community results in forming of tests and conclusions. This is the scientific process. Scientists, just like nature, oppose each other in essentially a *war* of ideas. While there is a great deal of collaboration, participating in a kind of argumentation is a key to science. This war of ideas requires knowledge of facts and principles in science. It requires an ability to engage in good argumentation. These qualities are essential for good science. This is often limited by misinformation given by the media and by pseudoscience beliefs.

## ■ KEY TERMS

aneurism
arteriosclerosis (coronary artery disease)
body of knowledge
biology
chemistry
control variable
deduction
dependent variable
dichotomy
diffusion
diuretics
empirical
experiment
generality
geology
germ theory
homeostasis
independent variable
induction

interrelated
inquiry
kinetic energy
Little Ice Age
method
natural world
Newton's laws of motion
osmoregulation
osmosis
pet hypothesis
physics
reasoning
relativity
repeatability
science
spontaneous generation
testable
vasodilators
weltanschauung (paradigm)

## ■ PROBLEMS

1. Define "science" to a friend. Give an example of **a.** an issue, **b.** a question, and **c.** a person in science.
2. In question 1 explain how your examples are important for you and our society.

3. Compare and contrast:
    a. Induction and Deduction
    b. Dependent and Independent Variables
    c. Pasteur and Pouchet
    d. Weltanschauung and Paradigm
    e. Scientific Method and Reasoning
    f. Repeatability and Empirical
4. Give an example of something that is NOT a characteristic of science.
5. Why is science important to you?
6. List three dichotomies in nature that you have observed.
7. Scientific Inquiry Exercise. Please read the research finding below and answer the related questions.

## THE MYSTERY OF "ASPARAGUS PEE"

Roughly 22% of the population experiences smelly urine several hours after consuming asparagus. Studies using gas chromatography/mass spectrometry techniques identified two compounds responsible for the smell of "asparagus pee." Researchers speculated that different people must metabolize asparagus differently. Thus, the other 78% of the population might be genetically different from those who experience the odorous pee. When they consume asparagus, it was speculated, that 78% do not produce the two compounds responsible for the foul odor of asparagus urine. The investigators collected and analyzed urine from a large number of randomly selected people who had consumed asparagus a few hours prior to urine collection. The scientists were very surprised to learn that 100% of the urine samples contained the two odor-causing compounds.

a. What observations did scientists make which stimulated their interest in conducting this investigation?
b. Describe the inductive inferences made through the research on asparagus pee. Decscribe the deductive inferences made through the research in asparagus pee.
c. What was the investigators' hypothesis?
d. What empirical data supported their hypothesis? What refuted the hypothesis? How did the results of their study change the weltanschauung of the scientists regarding asparagus pee?
e. Describe the experimental approach that the investigators employed to attempt to support their hypothesis. Was it a reasonable approach?
f. Given the fact that all of the subjects in the study had the smelly asparagus compound in their urine, what should be done to investigate why only 22% experience the smell? If you were the scientist(s), what would you do next to answer the question, given the experimental results obtained above? New hypothesis? New experiment? How would you test your new hypothesis?

***Did you know**...? It was later found that there is genetic variation at play—only 22% of people have the ability to smell the two compounds responsible for the odor of "asparagus pee".*[8] *Thus, everyone has the smelly urine after eating asparagus but only 22% of us have developed senses to detect it.*

# ■ REFERENCES

1. Lorenz, C. 1968. *A presentation to the Third Order of Saint Francis*. Bronx, NY: Capuchin Priest Order of Saint Francis.
2. Allen, G. and Baker, J. 2000. *Biology: Scientific processes and social issues*. Hoboken, NJ: John Wiley & Sons, p. 34.
3. Mackinder, H. J. 1887. On the scope and methods of geography, *Proceedings of the Royal Geographical Society and Monthly Record of Geography* 9(3):141–161. Quote is on p. 154.
4. Dobzhansky, T. 1973. Nothing makes sense except in the light of evolution, *The American Biology Teacher* 35:125–129.
5. Allen, G. and Baker, J. 2000. *Biology: Scientific processes and social issues*. Hoboken, NJ: John Wiley & Sons, pp. 61–66.
6. Huxley, T. H. 1868. "A liberal education, and where to find it," delivered to the South London Working Men's College. In A.P. Barr (Ed.). *The major prose of Thomas Henry Huxley* (pp. 205–223). Athens, GA: University of Georgia Press, 1997. Quote is on pp. 208–209.
7. Popper, K. R. 1959. *The logic of scientific discovery*. New York: Basik Books.
8. Reeher, J. 2012. The Mystery of Asparagus Pee (unpublished manuscript); White, R. 1975. Occurrence of s-methyl thioesters in urines of humans after they have eaten asparagus, *Science* 189:810–811.

# CHAPTER 2

# The Philosophy of Science

Introduction

**Early Philosophical Underpinnings**
Ancient Philosophy

**Introduction to Argumentation**

**Evaluating Scientific Research—Argumentation Analysis**

**Typology of Argumentation**

**Rhetorical vs. Dialogic Argumentation**

**Positivism**

**Modern Branching of the Philosophy of Science**
Post-Positivism
Relativism
Realism

## Introduction

The illustration in **Figure 2.1** on the next page depicts a checkerboard with a shadow formed by a cup. Look at boxes "A" and "B." Which is darker?

If you answered "A" you are wrong. You saw a lighter "B" because "B" is surrounded by darker boxes in the shadow of the cup and only *appears* lighter. Actually both boxes "A" and "B" are the same shades. Cut out two circles in paper and cover the checkerboard with only the "A" and "B" boxes visible. When only the two boxes are seen without the background, the shades should be the same. Their surroundings cloud the perception that one is darker than the other. Blocking out the surroundings allows you to compare the two boxes more objectively. Your answer therefore depends on your perspective. Science is very much like this; scientific conclusions are made with many factors influencing how they are both developed and implemented. The validity of scientific claims should be judged with an understanding of the societal and scientific

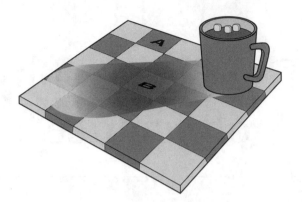

**Figure 2.1**   Checkerboard
Courtesy of Professor Edward H. Adelson

forces behind the conclusions. Scientific **arguments** should be evaluated using effective reasoning so that consumers of science can better interpret its truth and importance. Developing reasoning patterns is a main purpose of this chapter.

This reminds me of a quote a colleague in physics told me regarding my perspective concerning the world: "In the absence of what you are not, what you are isn't." In other words, if you are a short person but there are no tall people around you (the absence of what you are not), what you are (short) no longer is. You are only short if there are tall people around you. So too are scientific arguments, in general. Data in science are relative to the ideas and society surrounding them. Science is clouded by society much in the same way that the checkerboard squares cloud each other. Both the checkerboard riddle and scientific arguments can be clouded by a multiplicity of factors.

What clouds scientific thought? In many ways, society is a major factor in the form of existing scientific research and acceptable ways of thinking. If society deems a topic important, such as medical research, a push to study in that area will be made. However, most often money clouds the progress of science. Where does the money come from that is used to drive and even obscure science? Money is held and distributed by forces that are in power. Science progresses most quickly when the power structure decides to give it more attention and support. After all, during times of war science makes its most rapid progressions. The power structure spends more of the gross domestic product (GDP) in that direction.

Scientists may also hold *pet hypotheses*, which are guesses that the researcher holds onto too strongly and even unconsciously ignores evidence against. In the example of the Pouchet-Pasteur debate on spontaneous generation in the introductory chapter, Pouchet held on too strongly to his belief in his methodology and overlooked data and design. This may occur for many reasons, but often society and its generally accepted viewpoints cloud scientists in their development of hypotheses and even choice of topics. This may occur either consciously or unconsciously but it impedes progress nonetheless. Where does much of our research focus on in the U.S. today?

# Early Philosophical Underpinnings

Scientists challenged existing views of the natural world early in an era known as the scientific revolution (1540–1690). Two of the most important philosophers contributing to modern scientific methods are Sir Francis Bacon and Rene Descartes. Both were skeptics of prevailing thought; they rejected long-trusted authorities and ways of thinking. Instead they emphasized systematic, experimental-based *skepticism* about the world and based their theories on reason and mathematics.

Sir Francis Bacon (1561–1626) was an English politician who recognized the importance of science in benefiting commerce and industry. He was one of the first in the scientific revolution to popularize science, bringing it closer to the masses. Bacon was also a proponent of the *empirical method*, which is defined as a way of investigation that seeks to find knowledge objectively. Objectivity, he argued, required the collecting of data systematically. Studying phenomena, he felt, should proceed without the influence of ideas or prejudices held by a larger group or society. *Empiricism*, defined as the process of using the empirical method, is the basis of modern science. It views science as occurring in a vacuum, without the influence of society. Bacon believed that authority corrupted science and that it would hold back what science could give to humanity and industry.[1]

Lack of mathematical understanding, unfortunately, limited Bacon's contributions to science methods. His contemporary Rene Descartes (1596–1650), unlike Bacon, put his faith in mathematics. Descartes published *Discourse on Method,* which defended skepticism using the mathematical underpinnings of natural philosophy. He advocated use of deductive reasoning based on mathematics to question existing ideas and to form new conclusions in science. Modern science philosophy is based on *deductive methodology*, which states that conclusions need to logically flow from any premise. Descartes stated

> that inquiries should be directed, not to what others have thought, nor to what we ourselves conjecture, but to what we can clearly and perspicuously behold and with certainty deduce; for knowledge is not won in any other way.[2]

The way of Descartes was to question all authority from the ancient Greeks to the Bible. Any thought that did not include the process of deductive inference was rejected. Similar to Bacon, he assumed that the universe was based on set laws detached from human influence. His philosophy of science, called *Cartesian dualism*, defined the universe as matter that occupies space (Descartes terms this extension) and matter in motion: a separation of mind and body. Descartes also viewed the body as merely a machine relegated to the rules of the universe. He did, however, believe in the power of the mind, which gave him existence, and declared in the famous statement, "Cogito, ergo sum" translated to "I think, therefore I am." Descartes did, however, question all thought and posited that even his mind could not be certain of anything.

## Ancient Philosophy

Both philosophers departed from Aristotle's (the ancient Greek philosopher of the fourth century BC) view of thought, which was primarily observational and nonskeptical. The prevailing scientific thinking of the ancient to medieval world was ruled by Aristotle's views. It comprised a basic level of investigation as compared with the *empirical* methodology of the scientific process. While Aristotle had both inductive and deductive techniques for observing the natural world, he was primarily averse to skepticism about existing knowledge. His views rightly dominated scientific thought for millennia because they explained the observable world and made sense of so many phenomena. His profile is shown on the ancient coin in **Figure 2.2**.

**Figure 2.2**   Aristotle
© marekuliasz/ShutterStock, Inc.

However, modern science philosophy is based on higher levels of skepticism and argumentation, harkening back to Bacon and Descartes. Nonetheless, scientific thinking continues to begin with the kinds of observations used by Aristotle but is extended through skepticism.

## Introduction to Argumentation

Determining the way we gain an understanding of scientific knowledge and conversely, its limitations, is termed the *philosophy of science*. Any modern philosophy of science is based on an ability to effectively argue viewpoints and even counter viewpoints. Scientists always start with a premise, which is a *scientific statement*, in accordance with Descartes' deductive reasoning paradigm. These statements are attempts to explain scientific happenings. To illustrate, Galileo studied whether a 1-kilogram iron ball would hit the ground first when dropped from a building compared with a ½-kilogram iron ball (neglecting air resistance). Aristotle had claimed that heavier objects had greater speed when falling than lighter objects. Galileo made a scientific statement to the contrary, that speed of an object falling was independent of how heavy an object was. He, of course, tested it empirically by dropping both from a tower at the same time and measuring the time each took to hit the ground. He found Aristotle's idea, that there would be a difference between the two bodies, to be untrue. Both hit the ground at the same time, showing the speed of an object falling to Earth is independent of weight. Further investigation elucidated the exact acceleration at which a body falls to Earth, $9.8 \text{ m/sec}^2$.

A scientific statement is different from other statements in that it is testable. Galileo tested a statement and found that more investigation was needed. A scientific statement is thus actually an argument making some position. What is a *scientific argument*? It is a set of premises asserted in a certain way to establish the truth of a position (conclusion). It is a form of persuasion. Through studying the components and levels of arguments, one may effectively criticize and formulate one's own viewpoints and counter viewpoints.

## Evaluating Scientific Research—Argumentation Analysis

Science is complex and conclusions made by research are actually arguments about the extent to which something is true. Those arguments should be critically evaluated. For example, a University of California Los Angeles (UCLA) study was reported in October 2011 about the role bacteria play in causing pancreatic cancer, a dreaded disease in which only 5% of people survive after 5 years. The report shows that *Streptococcus mitis*, a type of oral bacteria, declines in numbers in the mouths of people with pancreatic cancer. Does the cancer fight the bacteria? Does the bacteria fight the cancer? Does

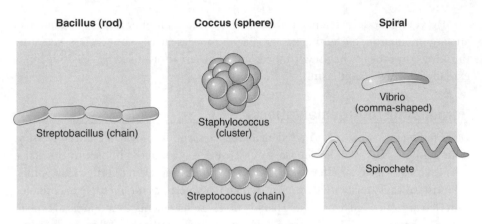

**Figure 2.3**   Variations in Bacterial Shape and Cell Arrangements

good tooth brushing reduce both cancer and bacteria? Or is it just coincidence? The relationship is difficult to deduce.

*Streptococcus* is a genus of bacteria that contributes to tooth decay and looks like chains of circles under the microscope, as shown in **Figure 2.3**. Each shape of bacteria is a specific identifier to the type of organism and also the disease associated with it. Unfortunately, media reports by several news outlets make the scientific argument that tooth decay is thus linked to pancreatic cancer and conclude that "good hygiene can avoid the terrors of getting the disease." While no real relationship between the *Streptococcus* and cancer can be drawn, reporting agencies make the erroneous conclusion.

This topic will be a useful example in analyzing arguments that make claims about a scientific issue. The following section will help the reader place scientific statements related to the topic into a typology (ranking scale) of argumentation that evaluates the strength of scientific arguments. I developed this typology in my study and use of argumentative strategy with certain premises based on King and Kitchener's reflective judgment model.[3] The chart is shown further in.

The lowest manner of argumentation in the typology is the Stage 1 observation level argumentation, which dates back to ancient Greek natural philosophers, such as Aristotle. They argued that explanation is based only on what is observable. In observational argumentation, one must derive from facts some principle that summarizes those pieces of data. Observations and conclusions at this level are acceptable only if they are obtained by the five senses: seen, touched, heard, smelled, or tasted. The abstract or logical is not considered. For example, Aristotle based his system of animal classification on this principle, dividing animals according to "their manner of life, their actions, and their dispositions using a system of categories involving paired opposites."[4] He did not seek to understand how the animals came to be or even how their internal structures worked, but simply classified and grouped information.

While observational argumentation is a starting point to determining truth for any scientific investigation, it is really the lowest type of reasoning. Thus, in this level of thinking, using our oral bacterial research example, if there is evidence that good hygiene can avoid the terrors of getting pancreatic cancer because bacteria cause the problem, then it must be so. The numbers support the results and a relationship is reported about *Streptococcus mitis* and tooth decay. In stage 1 argumentation, observations are the closest representation of truth and a relationship between tooth decay and pancreatic cancer is convincing enough. Observations are the sole source of truth of a scientific problem in this stage of reasoning.

Naturally, such conclusions may not necessarily be the full answer. In the next level, Stage 2 authority-determined argumentation, there is appeal to some authority or governing body for answers to scientific questions. Another authority-based individual may also determine scientific understanding. Thus, at Stage 2, if the scientific community accepts an idea, it must be the case. In the example of oral bacteria, because a scientific organization (UCLA) and a media outlet reported the conclusion, it must be true. Majority and popularity of thought prevails and the conclusion posited is accepted: "Good hygiene can avoid the terrors of getting pancreatic cancer." The media outlet also reported that many scientists believe this study to have merit; thus the conclusions must be true. This is usually how most media outlets present scientific breakthroughs to determine the importance.

Even scientists harken to the fact that most other scientists believe certain data and conclude that there is validity. This level of thought adds nothing to original research in science because it conforms to a kind of groupthink mentality that accepts the status quo. Science progresses, as discussed earlier, by breaking through the existing thought structure and developing new ideas. In addition, this often drives researchers to erroneously report studies in the interest of getting published and getting recognition.[5] Thus, the authority stage of reasoning can have deleterious effects on scientific progress.

In the Stage 3, belief-based level of argumentation, "belief" is incorporated as a higher level than simple acceptance of society-held thought or observations. While belief is not based on evidence, it is placed at a higher stage of reasoning because it at least contests another way of thought. It may adhere to its own "groupthink" mentality with other believers but at least a battle of ideas has started. At this level, there is a conflict between authority and an individual belief system. So, a belief that there is no truth to oral bacteria's role in cancer at least begins a questioning of the status quo. A belief that refutes brushing as prevention to pancreatic cancer at least starts a debate, which is inherent in scientific progress. Perhaps at this level, the arguer knows of someone who had excellent hygiene and still developed pancreatic cancer or simply has a hunch that it is too simple a relationship. Although the reasoning does not rely on data or solid studies, it at least uses some skepticism to question the authority.

Only in the next Stages 4 and 5, where evidence-based argumentative strategies are employed, is a substantiation of the premises introduced. This evidence may be used to refute accepted thinking about phenomena or support contemporary understanding. Use of evidence determines whether a true questioning of scientific facts begins. The level of questioning and strength of the evidence determines whether it is Stage 4 or 5. Gathering of evidence is a major shift from the lower three levels of reasoning. In the example of the oral bacteria study, upon closer inspection, data are evaluated and evidence is considered. This is a main characteristic of Stage 4 and 5 arguments to debate the validity of scientific statements. It is difficult to type the level of reasoning between Stages 4 and 5 but the strength of the evidence used to support a conclusion is the key factor in moving an argument from one stage to the next in classification.

Upon further evaluation of the data, the report is erroneous in its conclusion that hygiene helps prevent cancer because it does not match the data. People with pancreatic cancer actually have fewer *S. mitis* cells in their mouths so effective tooth brushing would have the opposite effect—increasing pancreatic cancer susceptibility. So why does the report make such a claim? It is probably because such a finding sounds good. Through making people feel empowered that they could somehow avoid pancreatic cancer by brushing, the report is positive, interesting, "headline grabbing," and makes sense to the public. However, it is an example of bad science in every way when applying the Stage 4 and 5 argumentative strategy.

In the final and highest Stages 6 and 7, the probability-based argumentative strategies use a variety of criteria to evaluate the strength of an argument. The following criteria are most important for evaluation: strength of the evidence, strength of the alternative evidence, and risks of error in accepting each of the evidence. To illustrate, Stages 6 and 7 use evidence in ways to bring forth a strong case for the truth of a particular premise. In short, the argument is strong because the evidence presented is based on an evaluation of the strength of the information. The conclusions in the oral bacteria example are not only wrong but no information is given by the media outlet about methodology, experimental design, or mathematics backing the data. In fact, upon further investigation at Stages 6 and 7, beyond the news media to other, more rigorous science journals, it appears that only 10 subjects were tested in the study and no real claims could be made. The sample size of 10 is very small and lacks the statistical power to support generalizations made by the media, which called for national testing of people's mouths for the disease.

Research reporting such as this bounces the public back and forth with weak research. A friend of mine gave up coffee for 3 months because of a report citing coffee and pancreatic cancer links. The death of Steve Jobs of Apple Computers and Hollywood star Patrick Swayze heightened people's awareness for pancreatic cancer, stimulating a rash of reporting over the past few years. With the knowledge that such claims will get attention, the media is all too willing to make headlines from the data. The media's inability to interpret the numbers behind the science demonstrates the need for improving the mathematics literacy underlying science literacy. A good argumentative strategy at Stage 6 or 7 would evaluate the best evidence possible and present that case. It would also look into the strength of the *counterarguments*. Counterarguments are defined as arguments against the original premise. This information can be used to refute or support a hypothesis, depending upon how it is presented and how it is placed within an argument. This determines the stage at which the argument is classified.

The highest stage 7 would evaluate the possibility of error in each of the pieces of data supporting the arguments. Such an argument would delineate the mathematics and research design strength and validity of the measures, to name a few considerations. It would study works cited about the topic and evaluate the opposing, alternative evidence of an argument to determine errors in both sides.

There are inherent probabilities of error in each of many research studies. First, mere relationships in the first argument do not imply cause. Perhaps there are other variables besides tooth brushing and pancreatic cancer. On the other hand there is also a pressure to produce positive results, which may lead to errors. The *Journal of Medical Ethics* reported a sevenfold increase in retractions in research related to mere errors in the period between 2004 and 2009.[6] Clearly, there is a need to question research on many levels including the inherent errors in their argumentation strategies. Ultimately though, a conclusion must be made, however plausible.

When determining the strength of any evidence in a scientific argument, there are a few very important points to consider. First, there needs to be an assessment of the relationships between the evidence and the scientific conclusion in question. In other words, does the conclusion match the data? Second, there is a raw tally of the number of points addressed versus the total number of possible points to address in an argument. That is, how inclusive is the argument in terms of how it covered the topic? Does each oral bacteria argument include all available data to defend the ideas or did some get left out? Finally, does the evidence supporting one's argument counteract the counterevidence? Being able to argue the other side's perspective is equally important in an argumentation strategy. Debunking the idea, for example, that hygiene is related to any disease suppresses doubt that tooth brushing is the culprit in pancreatic cancer (although oral bacteria have been implicated in many diseases ranging from heart disease to inflammations). It is difficult to judge the true strength of an argument beyond these points. However, these guidelines should allow for a most objective way to judge any scientific argument.

## Typology of Argumentation

The *Typology of Argumentation: Justification of Beliefs* is presented in **Figure** 2.4. It depicts the levels of argumentation, describes the stage, and assesses the level of certitude under the term "epistemology" of any scientific argument. The study of judging knowledge claims and determining what constitutes knowledge is termed *epistemology*. Epistemologically, each level bases its claims with a degree of certainty about their conclusions. The degree of certitude, oddly, decreases at higher levels of argumentation. There have been many scientists very sure of their data, as in Pouchet's case in the introductory chapter, who omitted important observations and conclusions in search for his or her "own" truth. The higher stages of argumentation are less secure in their knowledge because these levels recognize the uncertainty of scientific information. Good scientists realize, as Descartes did, that nothing is ever really all that certain.

Please be advised that each person is not stuck at one level of reasoning or another. Instead, we are all quite able to move from one level to another based on the topic or argument in question. A person may be very intellectual and well-reasoned, but his or her argumentation style may be at Stage 3 with regard to pancreatic cancer because he or she refuses to evaluate or consider alternate evidence and adhere to the belief-based Stage 3. At the same time, that same person can evaluate data on his or her own medical issue to make the best, reasoned decision on his or her health at the probability-based Stage 7. Developing reasoning patterns, however, takes practice and that is a major purpose of this text.

This typology can be interpreted in a number of ways. Jean Piaget (1896–1980), a famous cognitive researcher who studied how children form meaning, felt that people cannot move to more advanced reasoning until their brain develops enough to do so. He established *genetic epistemology*, which states that people cannot develop scientific thinking beyond the abilities that nature allows. He argued that children moved through stages of development to advance reasoning patterns; but nature was the ultimate determinant. Some people develop intellectually and others do not, according to Piaget. Thus, he contended, individuals were genetically locked into certain ways of thinking and natural development would determine the rate of advance in reasoning patterns.[7]

A more contemporary, accepted set of philosophers and educational researchers (e.g., Lev Vygotsky [1896–1934]) contend that humans are not locked into Piagetian reasoning levels and can develop cognitively through education and practice. Vygotsky studied children's play and speech patterns in the early part of the twentieth century. He found that there is a level of appropriate instruction based on a student's ability level. His work showed the improvements children demonstrate when given the right assistance at the appropriate levels. If given the right scaffolding or help, the student was shown to improve his or her scientific thinking skills. This level of instruction is known as the

| Stage No. | Stage Name | Descriptor | Epistemology |
|-----------|-----------|-----------|-------------|
| 1 | Observation | What is observed is truth | Certainty |
| 2 | Authority | Popularity or Authority Determines truth | Certainty |
| 3 | Belief | Conflict between Authority and Individual Belief System | Unsupported Uncertainty vs. Certainty |
| 4 | Evidence | Data and Valid Facts Relating to Conclusions | Approaching Uncertainty |
| 5 | Evidence | Stronger Data and Valid Facts Relating to Conclusions | Uncertainty |
| 6 | Probability of Truth | Evaluation of Evidence based on:<br>–strength of evidence<br>–strength of alternate evidence<br>–risks of error in accepting evidence | Uncertainty |
| 7 | Probability of Truth | Stronger Evaluation of Evidence based on:<br>–strength of evidence<br>–strength of alternate evidence<br>–risks of error in accepting evidence | Uncertainty |

Conclusions are Plausible Points of View:

Evidence strength depends on:

1. relation to question,
2. coherence (data matching conclusion),
3. counteracting alternative hypothesis,
4. number and quality of points address/total possible points
5. risks of error

**Figure 2.4**    Typology of Argumentation: Justification of Beliefs

"zone of proximal development."[8] This text agrees with Vygotsky's perspective and is dedicated to improving scientific reasoning through tapping into this zone.

In college classrooms, scientific ideas are presented and discussed, with the inherent uncertainty considered. Unfortunately, over 50% of undergraduates lack the advanced reasoning patterns described in our *Typology of Argumentation*.[9] Evidence suggests that these reasoning patterns can improve with teaching and learning strategies geared to this effect.[10]

## Rhetorical vs. Dialogic Argumentation

Determining exactly what a scientific argument is has produced the branch of philosophy known as *rhetoric*. This philosophy views the scientific method as "the faculty of observing and using the available means of persuasion."[11] This perspective advocates engaging in a form of reasoning called *rhetorical argumentation*. This reasoning method uses observation to argue a scientific point and develop a conclusion to persuade an audience of listeners. This is a traditional view of solving scientific problems.

There is a distinction between rhetoric and *logic*, made mostly in terms of the method of convincing people of one's beliefs. Rhetoric often relies on appeals to emotions, use of language, and methods of delivery. It encompasses the appearance of the argument more than its content or structure. On the other side, logic uses established rules of argumentation, avoids fallacies, and comprises only the strength of factual reasoning. Both rhetoric and logic are necessary components of any argumentation strategy.

In rhetorical argumentation, both logic and rhetoric are used to present different sides of an argument and are given by only one person, in a monologue, who determines the answer to the question. This is where the term "rhetorical argument" comes from. It is a way to persuade an audience without garnering audience input. The philosophy is that truth can be derived from one person's thought processes based on his or her observations. The answer can be given without input from any other people. Lectures, speeches, newspapers, internet articles, radio talk shows, and often TV are examples of rhetorical deliveries of arguments.

A more modern approach to arguing matters in science emphasizes the importance of communication and community, and not rhetoric, to search for truth. It uses a *dialectic process*, which delineates the different sides of an issue and brings the final, winning side forward for acceptance. Dialectics are sought by arguing in a back-and-forth manner until the true answer is revealed. It is the traditional form of debates in modern culture. This is also called *dialogic argumentation*. It includes an interfacing of a community of scientists to solve problems. Science philosophers such as Thomas Kuhn (1922–1996), Stephen Toulmin (1922–2009), and the current Götz Krummheuer are a part of the *dialogic* approach to scientific investigation. Instead of rhetorical arguments (that involve no one except the arguer), dialogic argumentation considers science a dialogue in which discussion leads to an exposing of contrasting ideas to facilitate reflective thinking. Krummheuer views science as a social phenomenon in which cooperating individuals try to adjust their intentions by verbally presenting rationales for their thoughts. There is discourse and truth lies in the product of a social event. The philosopher Toulmin states that science and scientific arguments are statements or decisions that are gradually supported. Kuhn argues in his writings that these interpersonal exchanges are essential to weigh information and make decisions with each person having expert input. Through this process a more convincing truth is produced, as shown in the cartoon in **Figure 2.5**.

Modern science involves, of course, both rhetorical and dialogic argumentation at points, but ultimately, it is a social process. To contribute to a community, scientific ideas need to be evaluated by other experts, criticized, modified, and retested. The community as a whole progresses through publication and networking of ideas, a concept that will be discussed further in other chapters. But in the minds of scientists, there is always a constant, rhetorical form of argumentation as described by Aristotle's classical approach to science.[12]

© RetroClipArt/ShutterStock, Inc.

**Figure 2.5**  Argumentation

## Positivism

A traditional view of science is that it uses a set of steps in uncovering the truth about a subject. The series of steps is called the *scientific method*. This method will be discussed in detail in another chapter but it is based on a branch of philosophy called *positivism*. Positivism is a system of philosophy basing knowledge solely on data gained from sense experience. It uses only scientific facts and their relation to each other and rejects speculation about the unobservable. Positivism *posits* a hypothesis about a phenomenon and seeks to either support or fail to support that hypothesis. As a hypothesis is supported through various investigations, it becomes more and more believable. This is how science facts form, according to positivism.

The conclusions formed from positivism are rather certain, with the power of the scientific method behind it. Positivism holds that larger, deeper meaning questions cannot be answered. "Why?" and "What will be?" are not able to be sensed through empirical and positivist methods and thus are not a part of scientific philosophy. They are moral and/or ethical questions that should not be tested.

## Modern Branching of the Philosophy of Science

### Post-Positivism

A major shift from the certainty of conclusions made by positivists resulted in a branch of philosophy known as *post-positivism*. Post-positivism (also called post-empiricism) is a scientific philosophy which argues that scientific knowledge is not based on solid facts but is always able to be changed. It criticizes positivism in a host of ways. Unlike positivists, post-positivists believe that human knowledge is weak and that the data derived from the scientific method can change readily as new knowledge is created.

There is no proof in science, though—only disproof of alternate working hypotheses, according to post-positivists. No hypothesis is ever proven because something could always be found to discredit it. Only a process of elimination of alternate ideas supports a hypothesis. This process is termed *falsification*. As discussed in the introductory chapter, this philosophy was first developed by a leading postpositivist, Karl Popper (1902–1994), who wrote many treatises on the philosophy of science.

Karl Popper advocated the view that there is only progress in science via falsification of existing knowledge.[13] Take the example of a medical problem such as a pain in the kidney. Do we know if it is a kidney stone or an infection or even cancer? No. It needs to be investigated and various possibilities need to be ruled out. Of course, it depends on the symptoms. Thus, through positivism and falsification of different hypotheses, a correct diagnosis may be made. Does that mean that it proves the diagnosis? No, again. A kidney stone may be detected by ultrasound and treated but then another cause of the pain could be lurking. This is why the public is so frustrated with their medical doctors.

Falsification leads to uncertainty and people hate uncertainty. Therefore people are frustrated by medical doctors who must use the scientific method.

Strictly following only the uncertainty of drawing conclusions within Popper's view of the positivist perspective would make the scientific community paralyzed. While we can accept that there is always an uncertainty in any scientific decision, conclusions must be made and be used. The positivist paradigm, which allows for solid conclusions, dominated from the scientific revolution through to twentieth century science philosophy.

All philosophies of science share a common theme. They deal with the changing branch of philosophy, *ontology*, which is the study of truth or reality. "What is truth?" is the most often cited question in any philosophy of science study or text. The goal of all of the philosophers of science, from the ancient Greeks (Aristotle, Socrates, and Plato) to the European revolutionists (Galileo, Copernicus, and Kepler) was to find truth. However, their ways of doing so differed.

Post-positivism incorporates the influence of society as well as unobservable phenomena. In this approach, science includes the unobservable and nontestable. Strict positivism rejects these. Post-positivism values what is seen but also what is yet to be determined. It is different from positivism in that it bridges forward from what is observed to use inferences on what might be. In the example earlier, the post-positivist would state that the kidney patient should be treated for his/her kidney stone if it is found. However, it recognizes that it may never be known what the true cause of the pain is. Maybe it is not the kidney stone causing the pain. Maybe it is something not yet discovered. It argues that conclusions may still be made on the unknown. Knowledge on the matter is not certain. Post-positivism is a philosophy that seeks to be useful to the public, in that it uses both the observable and unobservable to make decisions. It is open-minded enough to recognize the existence of unobserved parts of the picture. However, its goals are to draw useful but tentative conclusions to move forward within scientific study.

Post-positivist empiricism, nonetheless, does make a distinction. It recognizes that the observable is more important than the unobservable. However, because this perspective accepts that there are some unknown parts of a phenomenon, there is untapped potential to what data may yet be undiscovered about an issue. Either way, a conclusion is made without fully testing all parts of the picture. The patient still needs to be treated with the information given but knowledge may change. In the medical example, while no diagnosis is certain, the findings of the doctor should be used and the patient's kidney treated. If the pain persists, medical practice would require that more methodology is needed to investigate. Popper's words are warranted under this perspective but practical decisions still need to be made. The philosophy deemphasizes the importance of an absolute truth in the universe. It is at a higher level of argumentation in the typology than positivism. It recognizes the uncertainty of knowledge.

## Relativism

Two other new philosophical branches of science have recently emerged. First, a branch of philosophy, *relativism*, developed as a byproduct of dialogic argumentation. Relativism is the branch of philosophy that contends that all knowledge is true only insofar as it is compared with something else. In the example at the beginning of the chapter, the squares on the checkerboard were seen relative to the others around them. Truth was determined not as a certain and real fact but as related to how they appeared surrounded by the others. The squares were the same but that does not matter in relativism. If they appeared different to the observer, then they were indeed different.

In relativism, the role of society in scientific progress is emphasized. It argues that society creates various truths and changes only as the community's views change. Scientists discuss the existing data and theories but knowledge is fluid, with ideas changing as new information emerges and as society gives its inputs. Shifts in paradigms (models of thinking) are the way science changes, instead of a move toward some objective truth, as seen in positivism and post-positivism. Relativism states that science is more of a social function and less about a reflection of the workings of the natural world.[14] This philosophy in some ways downplays that objective reality may exist. It argues that people and not phenomena make reality. A pure version of relativism is accepted by very few scientists because the philosophy fails to take into account a reality that some objective and nonhuman truth is out there.[14]

## Realism

A third emerging philosophy of science is *realism*. Scientific realism is defined as a view that the world described by science and the scientific method is the real world, but dependent on society and opinion. It states that there exists a reality independent of human construct. However, it also asks the question, "What is the success of science in the world?" It seeks use and meaning in the world but reality and truth is dependent on that world.

It is, to some extent, a combination of both post-positivist empiricism and relativism. It searches for truth (or reality) based on both the observable and unobservable parts of a phenomenon, paralleling post-positivism. Conversely, it recognizes the relativist view that society intervenes to interpret that reality. In this way, realists view science as able to approximate what is true, while not exactly knowing everything.[15]

Realism is best described as a practical way of solving problems in the world. In the case of the kidney stone, the realist would argue that society is successful at treating problems if those treatments work. Treatment and a positive diagnosis is a success and when science correctly addresses real problems to find real solutions, realism is achieved. However, similar to other modern philosophies, realism accepts that there is some knowledge that is unknown. Perhaps the stone made some permanent damage leading to more infections for the patient in the future. There is no evidence for such a prediction in medicine to date, but perhaps it will be uncovered some day. Further research is necessary in this case. If society deems it important enough, work will be done to discover some future truth. If not, no progress will be made. The realist accepts that science cannot know everything but that it is still pretty good at getting to some degree of truth. It recognizes that society could change its paradigm of treatment of kidney stones, for example, to antibiotics if future professionals warrant it.[16] Again, the realist is rational in seeing the uncertainty in what potential science can accomplish. The realist accepts the failures and limitations of science as well. There are so many areas yet unknown in science, as discussed in the introductory chapter. Does acupuncture really work? Is there a real biochemical challenge to evolution? Will we cure cancer? Is string theory valid? Realism tries to integrate society's needs with these future possibilities.

Philosophical movements are complex and changing. It is unclear which direction the future of scientific philosophy will take but such shifts will determine the way science is done. Ways of understanding truth through science are emphasized by all three philosophical perspectives. The differences may well guide scientific thought in the twenty-first century.

# ■ KEY TERMS

| | |
|---|---|
| argument | philosophy of science |
| Cartesian dualism | positivism |
| counterargument | post-positivism |
| deductive methodology | rhetoric |
| dialectic process | rhetorical argumentation |
| dialogic argumentation | realism |
| empirical method | relativism |
| empiricism | scientific argument |
| epistemology | scientific method |
| falsification | scientific statement |
| genetic epistemology | skepticism |
| logic | *Streptococcus* |
| ontology | typology of argumentation |

# ■ PROBLEMS

1. What is a "pet hypothesis"?
2. Compare and contrast the following items:
   a. Rhetorical and Dialogic argumentation
   b. Piaget and Vygotsky
   c. Positivism and Relativism
   d. Realism and Positivism
3. Based on the *Typology of Argumentation*, identify which stage of reasoning the subject is in each of the given scenarios:
   a. A student is debating whether or not pandas are really an endangered species.
   b. A scientist is told that the polar ice cap is melting and accepts this.
   c. A sunbather looks at the sun and says it is yellow.
   d. A person looks at the moon and says it is the same size as the sun.
   e. A science student will not accept the embryological evidence of evolution.
   f. A patient is evaluating a hip operation's risk to determine whether to get it.
4. Evaluate the statement: "Science is a social phenomenon." Show one way it is true and one way it is not true.
5. Write a one-page reflection on a scientific argument you engaged in during the past 24 hours.

# ■ REFERENCES

1. Sherman, D. and Salisbury, J. 2011. *The West in the World* (pp. 452–454). New York: McGraw-Hill.
2. Ibid, p. 453.
3. King, P. M. and Kitchener, K. S. 1994. Developing reflective judgment (pp. 14–16). San Francisco: Jossey-Bass.
4. Leithauser, G. and Bell, M. 1987. *The world of science: An anthology for writers.* New York: Holt, Rinehart and Winston.
5. Steen, R. G. 2011. Retractions in medical ethics: How many patients are put at risk by flawed medical research? *Journal of Medical Ethics* 37:688–692.
6. Ibid.
7. Piaget, J. 1970. *Science of education and the psychology of the child.* New York: Orion Press.
8. Allen, R. 1981. Intellectual development and the understanding of science: Applications of William Perry's theory to science teaching. *Journal of College Science Teaching* 12:94–97.
9. Hofer, B. and Pintrich, P. 1997. The developments of epistemological theories: Beliefs about knowledge and knowing and their relation to learning. *Review of Educational Research* 67(1):88–140.

10. Daempfle, P. 2006. The effects of instructional approaches on the improvement of reasoning in introductory college biology: A quantitative review of research. *The Journal of College Biology Teaching: Bioscene* 32(4):22–32.

11. Leithauser, G. and Bell, M. 1987. *The world of science: An anthology for writers.* New York: Holt, Rinehart and Winston.

12. Kuhn, T. S. 1970. *The structures of scientific revolutions,* 2nd ed. Chicago: University of Chicago Press.

13. Schick, T., Jr. and Vaughn, L. 1995. *How to think about weird things: Critical thinking for a new age.* Mountain View, CA: Mayfield Publishing Company.

14. Lee, J. 2000. *The scientific endeavor: A primer on scientific principles and practice* (pp. 23–24). San Francisco: Addison Wesley Longman, Inc.

15. Ibid, p. 22.

16. Psillos, S. 1999. *Scientific realism: How science tracks truth.* London: Routledge.

# CHAPTER 3
# Scientific Research

## Introduction

This chapter shows the intricate nature of research in science. While there are certain, specified steps of a scientific method for conducting an investigation, research is more complex. Actual research involves many factors and revisiting of stages after finding out results of an investigation. As described in another chapter, for example, formulating a medical diagnosis requires changes of strategy and renewing "hunches" about symptoms. In science, there is always a need for reasoning about ideas, both inductively and deductively, to successfully conduct the investigational research.

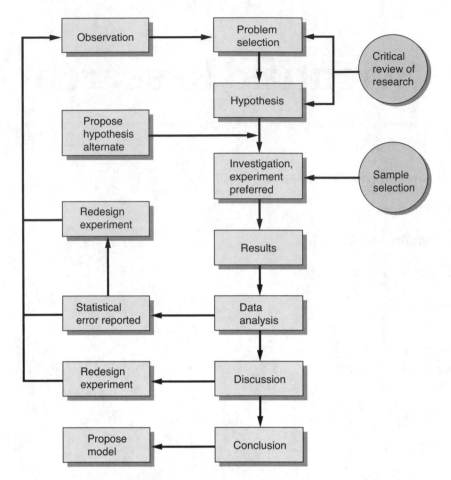

**Figure 3.1** Steps of the Scientific Method

The general steps of the *scientific method* are given in **Figure 3.1**. The scientific method is defined as a set of systematic procedures for organized observation and theory building. It is the way in which scientific knowledge is gained.

## Observation and Selection of a Problem

Studying the real world and its phenomena naturally leads human beings to ask about some of life's unanswered questions. Children constantly ask about natural phenomena: "Why is the sky blue?" "Why is the moon following us?" A child's curiosity is the starting point for any good scientist. In fact, a child's education is learning about what is known and asking the right research questions about what is not known.

This natural inquisitiveness drives science research through our society and our lives. While we think we know that the moon does not "really" follow us, why and how is the moon moving "with us"? There are many guesses and there is a lot we don't know about the moon. The answers lie in our ability to observe the obvious. The first step in scientific research is to make *observations*.

Then, a *selection* of a research topic to study leads to a background review of the literature about the **research problem**. A research problem is the object or process that needs to be studied. In the case of lunar applications, the existing knowledge needs to be explored to determine where society left off regarding moon chemistry and physics. A **critical review** of the literature requires that the topic be looked at from every angle. "Critical" means just that—to evaluate inconsistencies in reported research results, find flaws in their designs, identify alternative routes to study topics and critique the mathematical findings. To be a scientist is to be critical of existing knowledge. In this way, new perspectives to research questions are studied and new discoveries are made.

In 1985, a research study overturned the way peptic (stomach) ulcers are studied and treated. A scientist in Australia, Barry Marshall, proposed that the bacterium *Helicobacter pylori* caused, and was not a result of, gastric ulcers. Existing research said that these ulcers were brought about by stress and diet and that an ulcer may contain bacteria but only because the ulceration was a place conducive for bacterial growth. In other words, ulcers were not caused by bacteria.

The reluctance of his colleagues to accept the idea that *H. pylori* caused ulcers provoked Marshall to act. In an unprecedented move, the scientist made himself a human guinea pig and ingested a vial of active *Helicobacter pylori*. Marshall recalled in a recent interview:

> Those were frustrating times for me. Most of the experts believed that the presence of *H. pylori* in those who turned up with ulcer problems was just a coincidence. I planned to give myself an ulcer, then treat myself, to prove that *H. pylori* can be a pathogen in normal people. I thought about it for a few weeks, then decided to just do it. Luckily, I only developed a temporary infection.[1]

To the surprise of the scientific community, he formed a number of irritations in his stomach. This showed that his hunch was right—that bacteria did cause gastric ulceration. Whether or not his methods were advised or ethical is another matter.

## Hypotheses

The above example shows how after a research problem is identified a **hypothesis**, or educated guess, is developed and tested. A hypothesis is defined as a possible explanation for a natural phenomenon. In the case of the ulcer, the hypothesis was that *H. pylori* caused gastric ulcerations. It is a conceptualization of the research problem relating the facts to a testable question. The hypothesis should make sense and be based on reasoned research.

There are some general guidelines for evaluating a hypothesis. First, it must be *empirically testable*, meaning the results must be measurable. Second, it must be *logical*. That is, it cannot contradict itself. Third, it must address a question about a *natural phenomenon*. It cannot explore pseudosciences that are immeasurable and within the realm of belief. UFOs and acupuncture can be studied by suitable hypotheses, for example, but must reflect reality by giving measurable results. Fourth, it should seek to explain and *further science*. A hypothesis is an unchecked idea and is only a starting point in the scientific method. Tests of the idea are the true measure of the value of any hypothesis. A hypothesis is only a guess and as such is very subjective.

## Scientific Investigations

There are several approaches that can be taken to investigate natural phenomena. Naturalistic observations, scientific modeling, experimental and nonexperimental research will be discussed as possible methodologies. *Naturalistic observations* are studies describing how something works in a natural, real setting. Describing a chipmunk's behavior in a forest is a naturalistic observation. It gives the best picture of how the animal works in its normal everyday life. Opposite of this, *scientific modeling* predicts results based on simulations of the real world. Models are developed because there is no way to observe it in nature. An example would be a model predicting a hurricane's course. Because the event has not happened yet, it needs to follow predictable patterns based on certain assumptions about the way things usually work.

*Nonexperimental research* studies general features of events, categories, or places. It is observational and does not seek to intervene in what is being observed. This approach is useful, for example, to map a lake bottom or count the number of fish species within the lake. It is informative and lays the ground work for future studies. These sets of investigation methods will be treated in more depth later in the chapter. However, only the experimental design, described in the following section, which limits subjectivity and bias, is acceptable within the scientific method.

## Experimentation

The plan for an *experiment* should be detailed and organized and account for as many intervening problems as is possible. What is an experiment? It is an investigation that is organized in a certain way. As discussed in other chapters, it has a *control* and an *experimental* group(s) (that is treated with the studied variable), and *independent* and *dependent variables*. Again, an independent variable is controlled by researchers and a dependent variable is the result of a study. In this way, the influence of the independent variable on the dependent variable can be best determined. Through keeping as many variables as possible constant (not changing) during the experiment, the results are more robust. That is, the dependent variable truly "depends" only on the changes of the independent variable.

Consider an experiment in which reaction time is tested according to age. Our hypothesis states that the time it takes to react to stimuli declines as we age. A simple experiment can be set up in which different age groups are required to "catch" a ruler dropped vertically within the grasp of their hand. The researcher may record how many centimeters of drop it took for each subject to catch the ruler.

What safeguards could be put into place to assure that the experiment is controlled? The same procedure for dropping the ruler, the same conditions of light and temperature for all subjects, and perhaps even the same tone of voice should be used while the researcher drops the ruler. Clearly, there are many variables to control while conducting a sound experiment. Otherwise, *extraneous variables* may influence the results of the experiment. Extraneous variables are those that influence the results of an experiment but are not meant to. They are not being studied but may interfere with the influence of the independent variable on the dependent variable. In the reaction time experiment, lighting conditions are one example of an extraneous variable that could affect eyesight and, therefore, reaction time. Careful preparation before an experiment is conducted can minimize the effects of extraneous variables and make the results of the study more valid. This is termed "controlling" the experiment. **Figure 3.2** shows a good experimental design.

## Sampling

Usually, it is not possible to study whole populations. In order to study a whole population effectively, a sample or smaller subgroup of the population should be chosen carefully. In an analysis of cholesterol levels in the U.S., an individual cholesterol measurement taken for each member of the nation's population is not economically feasible. Also, studying a single subject's cholesterol test does not provide results applicable to the general population. Therefore, thorough and careful selection of a sample is important. Perhaps taking into account different age groups, fitness levels, and weights would provide a population sample with blood test results more reflective of the overall population.

Many studies do not use the term *subject*, which is usually reserved for humans, but instead refer to a *sample group* indicating human and nonhuman units for study. Some sample groups are even geographical regions. For example, when looking at mercury deposition to New Jersey lake sediments, the sample group is a set of

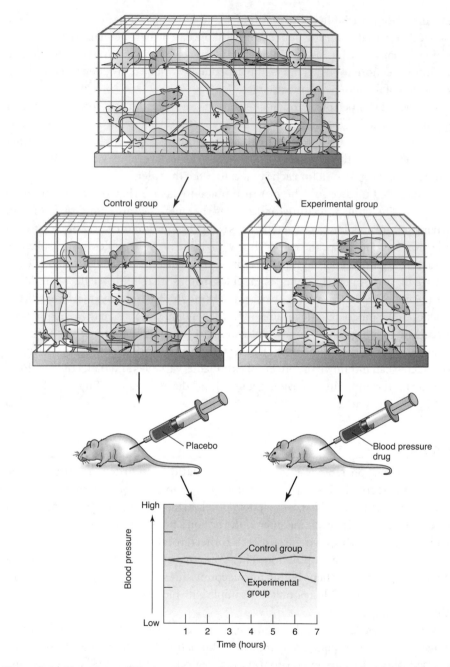

**Figure 3.2** Experimental Design: Rats are placed into two groups: control and experimental. Both groups are treated the same except for one variable. In this case, the experimental group is given a blood pressure lowering drug. The control group is given an injection of physiologic saline, the solution the drug is dissolved in (and should have no effect).

measurements from different locations from which soil and sediment samples are collected for chemical analyses. The sample group should always be representative of the larger population.

## Bias

The results of any experiment may be flawed due to various types of *bias*. Bias occurs whenever objectivity is compromised. Bias is a mental leaning or inclination, partiality, prejudice, or bent to a researched problem. It gets in the way of obtaining the truth about a phenomenon.

### Sampling Bias

*Sampling bias* occurs when the sample is somehow not representative of the general population. In the above example, bias would occur if samples were collected only from urban areas of New Jersey. The right set of sample areas (e.g., urban, rural, near industrial sources, or in small towns) need to be selected to create an accurate view of mercury deposition representative of the entire state. Sampling only rural regions might result in lower reported rates of deposition, not truly representative of the state as a whole. Thus, careful selection of subjects or sample groups is vital to maintaining the generalizability of any experiment. It is usually most representative to choose random subjects from a whole population. This corrects for possible sampling error or bias.

### Psychological Bias

Natural studies present another type of bias problem. As stated earlier, the type of study looking at real life phenomena in nature is termed *naturalistic observation*. It occurs when the investigator gathers data in as natural a setting as is possible. Observations in nature can become unnatural. A chipmunk will not act naturally if it knows it is being observed. In fact, humans often do not act naturally when they are observed. An effect on results due to the study being conducted is termed *psychological bias*. When limiting this bias, a *blinded study* is conducted so that the subject does not know it is being observed. In medicine, often a *placebo* (nonpotent) is given in place of a drug, without the subject knowing which it is taking.

### Experimenter Bias

Conversely, the experimenter could treat subjects differently based on the groups they are placed in. This form of *experimenter bias* can lead to an experimenter unconsciously reporting results more favorable to their particular viewpoints. Keeping secret the type of treatment the subject is getting reduces this form of bias. This is accomplished by a *double-blinded* study, in which both the subject and the experimenter do not know in which group the subject is placed.

# Data Analysis

Information collected from any study needs to be analyzed systematically and objectively to form conclusions. A variety of research methods may be used to evaluate data. Data analysis may be either quantitative or qualitative in nature.

## Quantitative Analysis

*Quantitative analysis* is defined as the reporting and use of data that are numerical in scope. It is the more traditional scientific analysis system and depends on numbers to find patterns and draw conclusions in the results of a study. The main benefit of quantitative analysis is that it reveals relationships between variables that allow generalization of the results to a larger population. Often, this requires a large number of individuals or units in the sample. To illustrate, consider observing the effects of a drug on rheumatic hand joints. Large numbers of individuals would need to be sampled to allow for adequate statistical analyses. Statistics is the study of the collection, organization, analysis, and interpretation of data. For example, it would help scientists determine if the effects of a drug on improving joint damage is significant. If so, it can be recommended for arthritis patients. Quantitative data analysis is the focus of another chapter. Without math, science has little power to make recommendations or generalizations. Quantitative analysis is what separates science from the many forms of pseudoscience.

## Qualitative Analysis

On the other hand, *qualitative data* analyses have fundamental importance in adding richness of detail to scientific studies. *Qualitative analysis* is defined as the reporting and use of data that are nonnumerical in scope. It usually studies very few subjects or data pieces but looks at those in greater depth than quantitative research. These studies generally have smaller sample sizes and thus lack generalizability. To illustrate, a single study looking at an individual's rheumatic hand joint and the psychological effects of drug therapy on that person can provide important insights into the drug's overall usefulness. It may look into the patient's personal life, journals and diaries, and effects on the family life to illustrate.

Quantitatively analyzed, an x-ray of the joint may show marked improvement in the rheumatic hand after taking the drug, placing a subject in the "success" group. Qualitatively, however, perhaps the person suffered terrible sleep deprivation and depression resulting in a diminished quality of life. Only through a more in-depth approach would these kinds of results be reported and considered. This is the strength of the qualitative analysis.

Naturalistic observations, or "watching" the subject or setting in its natural proceedings, generally use the qualitative type data analysis. Keeping a journal, taking copious notes, and interviewing are methods of qualitative data collection. For human subjects, conducting focus groups, in which subjects meet in an organized manner to discuss certain topics, allows the researcher to observe in a seminatural setting. In this way, subject interactions and sometimes confrontation may be studied. Data from qualitative studies have a systematical method of coding for detecting patterns and making conclusions.

## Results

Basic analysis of the data and reporting of the numbers are placed in a *results* section of any investigation. Reporting results is meant to be objective, statistical and numbers-driven. Results are presented in a pithy manner, with little room for elaboration. The chapter on mathematical inquiry will discuss the detailed, quantitative results analysis, which is placed in a results section of a study. Results are never more than straight reporting of the facts. Even in qualitative data reporting, results are trend-based and point to conclusions.

However, there is often pressure on the scientist to present *positive results*. That is, results that support the hypothesis. It is often easier to publish a study with a supported hypothesis. Publication of positive results may come with rewards of prestige and money, whereas negative results (refuting original hypotheses) have less impact. Proper reporting of negative results is vital to science, because it ends or limits false leads and drives research into different directions. Falsely representing data and results is highly unethical, as will be discussed in the chapter on scientific integrity. Results should always be objective and unassuming so that readers can draw their own conclusions, possibly conclusions overlooked by the original researcher. In this way, science progresses more successfully with community input.

## Discussion

The results are *objective* and the hypothesis is *subjective* and so the *discussion* returns the investigation back to the subjective realm. The discussion interprets the data, explaining it based on the accepted literature as well as the intuitions of the scientist. It is the part of the investigation that is most creative and perhaps even speculative. However, it must be embedded in valid information; the information derived from the results section. The discussion answers questions as to where future research should head, confirms or refutes what is already known, makes alternate explanations for the data, and essentially shows the thinking of the actual scientist.[2] Conclusions are drawn from the analysis of the results in the discussion section.

## Scientific Modeling

When actual phenomena are difficult to study, a model is formed about the way something works. As stated earlier in the chapter, a scientific model is a simulation of the real world. A model may derive from an investigation or series of studies. Models make predictions about how something will work by extrapolating from the artificial condition to the real world. Global climate models give predictions about how things will be in the future given current atmospheric conditions. We have no way to see into the future. Thus, the predictions are based on mathematical or artificial situations.

Cancer is an abnormal and uncontrollable production of cells, forming growths called tumors. Tumors may spread to other parts of the body and lead to death. For the past 2,000 years, a group of people in Lin Xian, China, about 250 miles (402 km) south of Beijing, have been dying at high rates (one in four) from cancer of the esophagus, which is the muscular tube moving food from the throat to the stomach. Scientists investigated the Lin Xian area and discovered that the food being grown by the people was low in molybdenum, a soil nutrient for plants needed in small amounts. This caused the crops to concentrate nitrates from the soil to make up for the low molybdenum levels and reduced the vitamin C produced by the plants. Nitrates in the plants were being converted to nitrites in the stomachs of residents; and nitrites are linked to various digestive cancers, including esophageal. It is also known that low levels of vitamin C promote the conversion of nitrates to nitrites. When people were given vitamin C tablets, their production of nitrites dropped. Also, to help reverse the crop production of nitrates, the villagers coat their corn and wheat seeds with molybdenum. As a result, nitrite levels in vegetables have dropped 40% and vitamin C levels have risen 25%. The linked relationships to the villagers' cancer problem are an example of a model, based on scientific investigation, which reveals truth about a phenomenon without actually manipulating variables. It is a model for the workings of the Lin Xian esophageal cancer phenomenon, as shown in **Figure 3.3**. It makes only predictions for how the cancer develops based on the data and model derived. However, it is too early to tell whether or not cancer rates will actually change.[3]

Modeling is an important part of science but is limited due to the removal of real world manipulation of variables. After all, science can cure cancer in mice with anti-angiogenic drugs but in humans the model becomes sketchier. Mice are easier to study because, ethically, more can be done to reduce experimenter errors and eliminate extraneous variables. But in humans, the treatments for cancer don't always work the same. While mice have some biological similarities to humans, they are quite different. We cannot ethically control all aspects of human subjects' lives to eliminate extraneous variables. One group may be eating a different diet or be genetically unlike the other, thus conflating the results of the control vs. experimental group. The models developed from research on mice are sometimes frustrating to the medical community when applied to humans. Despite this, models are important for guiding research. The quote below by Kenneth Paigen expresses this.

> People have a basic misunderstanding of the mouse. They get upset that the exact pathophysiology might be different, but a mouse is not a total mimic. A mouse is a discovery tool, a device, to understand the molecular pathways

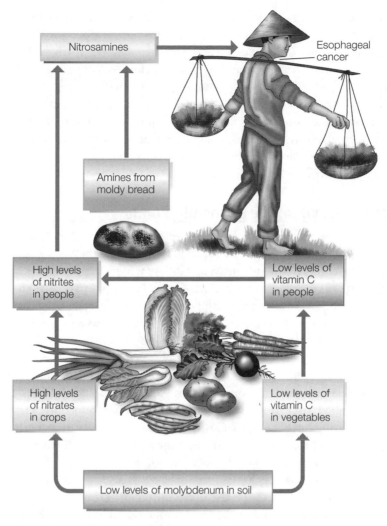

**Figure 3.3**   Tracking a Killer

that underlie disease processes. You can't do a lot of types of experiments on humans. You can't order people to mate... We have to turn to models.[4]

## Nonexperimental Research

You may have noticed that I have been careful not to use the word "experiment" capriciously in this chapter. A better term is "study" or even "investigation," if the research does not conform to the experimental design. Many studies are *nonexperimental*.

Nonexperimental studies are descriptive: They describe a natural phenomenon. They do not get set up by isolating a variable and having a control group. Nonexperimental studies and investigations are not meant to test predictions, as seen in hypothesis testing and experimentation. Exploring the coast of Iceland for levels of pesticides, describing the Martian soil and topography, and counting species in the Antarctic Peninsula are examples of nonexperimental research. These studies are an inductive gathering of information and they are science. Each furthers our understanding of the universe. Nonexperimental research is foundational in the history of science as will be discussed in other chapters. It is how science began in our history and is still the beginning phases of any scientific investigation, both experimental and nonexperimental.[5]

## Analyzing Research: The Incredible Potato

The best way to understand this chapter on scientific research is to read about and analyze an investigation. Consider the study in **Box 3.1** from the *New England Journal of Medicine* reporting a nutritional analysis of different foods in peoples' diets and their resulting weight gain or loss over a period of time. Please pay particular attention to the parts of scientific investigations that give it the strength to make valid conclusions.

In consideration of the last question asked, a host of news agencies reported the results to the public. A *Wall Street Journal* article appeared within a week of the study entitled, "You Say Potato, Scale Says Uh-Oh," which implicated potatoes as a major cause of weight gain and cited in its front cover photo that "boiled, baked or mashed potatoes correlate with a 0.57 pound weight gain" per year for subjects studied.[6] The article implied that even the nonfried potatoes are bad for people. Surprisingly, when investigating further outside of the article facts, the nutritional information on potatoes is very favorable, indicating that they are very healthy to eat. Consider that a boiled potato, cooked in its skin without salt and butter has only 68 calories; is very low in saturated fat, cholesterol, and salt; and is also a good source of vitamin B6, potassium, and Vitamin C. It is the added butter and cream in mashed potatoes or the sour cream and butter on baked and boiled potatoes that likely leads to the weight gain cited.

However, the article omits these extraneous variables co-occurring with the potato-weight gain link. What is the motivation? Is it merely oversight or is it part of a larger groupthink to eliminate carbohydrates from diets? Did this give the newspapers fodder for their anti-carb campaign? Many other news outlets reported the same misleading information about the results of the study, which indicates that it is not a lone error or biased reporter but part of a larger movement. As shown in the example above, the media's reporting of scientific information is often truncated and the public may be easily misled about scientific research results.

## BOX 3.1    ARTICLE ANALYSIS: THE INCREDIBLE POTATO

**Title: Changes in Diet and Lifestyle and Long-Term Weight Gain in Women and Men, an adapted summary of research by:**

**Background**
While eating less and exercising more is advised, eating different types of foods may contribute to weight gain and weight loss. Specific dietary food choices may be more important than total calories in determining weight and fitness.

**Methods**
A sample of 120,877 U.S. women and men were studied in terms of their dietary intakes. Samples were separated into three distinct cohorts. Subjects were free from chronic diseases and not obese at the baseline of the study. Follow-up periods were conducted at four-year intervals to evaluate lifestyle factors with weight changes. Adjustments to the data were performed in terms of age, body-mass index, as well as other factors. Various statistical tests were performed to evaluate effects of sex and grouping of the subjects to determine these influences.

**Results**
The following results were found in each of the four-year period intervals: Subjects gained an average of 3.35 lb, overall (5th to 95th percentile, −4.1 to 12.4); Dietary factors most strongly associated with weight gain were: intake of potato chips (1.69 lb), potatoes (1.28 lb), sugar-sweetened beverages (1.00 lb), unprocessed red meats (0.95 lb), and processed meats (0.93 lb); Dietary factors most strongly associated with weight gain were: intake of vegetables (−0.22 lb), whole grains (−0.37 lb), fruits (−0.49 lb), nuts (−0.57 lb), and yogurt (−0.82 lb); General lifestyle factors were also associated with weight change: increased levels of physical activity (−1.76 lb across quintiles); alcohol use (0.41 lb per drink per day), smoking (new quitters, 5.17 lb; former smokers, 0.14 lb), sleep (more weight gain with <6 or >8 hours of sleep), and television watching (0.31 lb per hour per day).

**Conclusions**
Specific dietary and lifestyle factors are independently associated with long-term weight gain, with a substantial aggregate effect and implications for strategies to prevent obesity. (Funded by the National Institutes of Health and others.)

In the investigation in Box 3.1:

1. What is the independent variable(s)?
2. What is the dependent variable(s)?
3. Describe the control group(s)?
4. What kind of evidence was collected in the study?
5. What is the hypothesis of the researchers?
6. Is this a quantitative or qualitative study? Why?
7. What is the conclusion of the researchers?
8. Are there any extraneous variables?
9. Do you see flaws in the research? Is there bias in the reporting?

*Sources:* Dariush Mozaffarian, M.D., Dr.P.H., Tao Hao, M.P.H., Eric B. Rimm, Sc.D., Walter C. Willett, M.D., Dr.P.H., and Frank B. Hu, M.D., Ph.D. *N Engl J Med* 2011; 364:2392–2404, June 23, 2011

# ■ KEY TERMS

bias

blinded study

critical review (of research)

discussion

double-blinded study

experiment

experimenter bias

extraneous variable

*Helicobacter pylori*

hypothesis

naturalistic observation

nonexperimental research

objective

placebo

positive results

psychological bias

qualitative analysis

quantitative analysis

research problem

results

sample group

sampling bias

scientific method

scientific modeling

subjective

# ■ PROBLEMS

1. Describe the process by which a research study is selected.
2. "A hypothesis is only an educated guess." Discuss the extent to which this statement is true.
3. Compare the three types of bias. How does awareness reduce bias?
4. What is an extraneous variable? How does it limit statistical power and experimental design?
5. Compare and contrast the following set of terms (give one way the terms are the same and one way the terms are different):
   a. Quantitative and Qualitative Research
   b. Results and Discussion
   c. Nonexperimental and Experimental Research
   d. Control and Experimental Group
6. Which is the most objective in an investigation and why: Results, Hypothesis, or Discussion?
7. What is a scientific model? What are its benefits and limitations in scientific progress?
8. How is nonexperimental research important to scientific progress?

# ■ REFERENCES

1.  National Institutes of Health: Office of Science Education. Available online at http://science. education.nih.gov/home2.nsf/Educational+ResourcesResource+FormatsOnline+Resources+ High+School/928BAB9A176A71B585256CCD00634489. Accessed July 6, 2012.
2.  Keppel, G., Saufley, W., and Tokunaga, H. 1992. *Introduction to design and analysis: A student's handbook*, 2nd ed. New York: W.H. Freeman and Company.
3.  Chiras, D. 2006. *Environmental Science*, 9th edition, Burlington, MA: Jones and Bartlett Learning.
4.  Lewis, R. 1998. How well do mice model humans? *The Scientist* 12 (21):10–11.
5.  Lee, J. 2000. *The scientific endeavor: A primer on scientific principles and practice*, San Francisco: Addison Wesley Longman, Inc.
6.  Hobson, K. 2011. *Wall Street Journal*, You Say Potato, Scale says Uh-Oh, June 23, 2011.

# CHAPTER 4

# Math Gives Science Power

"Philosophy is written in this grand book, the universe… It is written in the language of mathematics, and its characters are triangles, circles, and other geometric figures without which it is humanly impossible to understand a single word of it."

Galileo Galilei (1564–1642)[1]

## Introduction

Without mathematical analyses, scientific research is powerless. As expressed in the quote above by Galileo, the Italian natural philosopher of the 1600s, mathematics is the language of science. When an investigation is conducted, the results need to mathematically show that something actually happened or that a conclusion can really be supported. If, for example, more heart patients fare better with *angioplasties* (minimally invasive procedures that open up clogged arteries) rather than heart bypass surgery, the scientific community needs to know how much better. If the research shows higher survival in one group of patients than another, "How much?" and "Is it really enough of a difference?" are the next two questions that should be asked.

Many times a study is presented and even published claiming a difference between two study groups but no mathematical analyses are given. One should immediately disregard these findings as inadequate. When no numbers are given, the next question should be, "What are the authors hiding?" Usually, the newspapers, TV, or internet claim that there is limited space or time to give a full picture of what is happening when they announce conclusions of scientific research. While there is some truth to this, important science-based decisions are made from information dispersed by the media. This chapter will give readers the necessary mathematical background to more carefully consider any research study they come across.

Consider the full article presented below from the newspaper, *USA Today*, discussing the benefits of two types of angioplasties used to open occluded (clogged) arteries in atherosclerotic patients:

### Coated artery stent less likely to reclog

"An antibiotic-coated stent developed by Johnson & Johnson prevented reclogging of coronary arteries during a seven-month period, researchers said Tuesday. Stents, which are tiny metal tubes or grids used to prop open clogged arteries, often narrow from scar tissue that develops after insertion of the devices. Different companies are coating stents with various drugs in an effort to block growth of the scar tissue and remove the need for surgeons to redo the procedure."[2]

The article then concludes with a statement from a Johnson and Johnson spokesperson discussing only the cost of the procedure: $2,100 to $2,200 for the coated device versus $1,200 for a bare metal stent.

There are many criticisms that can be made about the way in which the study was announced by *USA Today* in the above article. Based on the guidelines from the chapter on scientific research for what constitutes good research, the newspaper's reporting was considerably unscientific. First, no data are given regarding the number of patients studied, sampling methods, or long-term follow-up on survival and quality of life of the patients. Second, there is no information on the experimental design, selection of subjects, or the control group. Third, there is almost no science explaining the procedure or complications. Fourth, and perhaps the most disturbing aspect of the newspaper article, is the lack of any statistical analyses. The reader is left without knowing if the results are really mathematically significant. The only numbers given are the price of the stenting procedure and the only science about the procedure is the general biology of the technology. This article was written in September 2001; a decade later, research indicates the drug-coated stenting procedure described may be linked to strokes and other complications. Its use and longer-term survival rates are now in question.

Many people influenced by reading this article and others like it may have opted for the drug-coated stents. If they were not provided with proper statistics, were they misled by the presentation of the article writer(s)? Did omission of the important mathematical bases for making conclusions mislead readers to a false sense of security about the new procedure? It is clear that a familiarity with statistical analyses is essential to a fuller understanding of scientific research.

Another defense of the way the news reported the stenting procedure may lie in the inability of the general public to understand mathematics and the statistics needed for properly evaluating a scientific investigation. This lack of the mathematical aspect of scientific literacy is the core problem for such types of reporting. If the public doesn't understand the math anyway, why waste paper and print it? Also, perhaps the public doesn't want to hear about the math behind the science. Nonetheless, the resulting lack of knowledge leads to poorly informed decision making on the part of the public and disempowers our populace. A main objective of this text is to highlight the importance of the mathematics component of science literacy in a properly functioning society. It is therefore vital that a working knowledge of statistics be treated in the next section. The aim here is to empower the reader to more carefully consider and question science investigations and their claims.

## Hypotheses and Statistical Error

Statistical procedures are used to determine whether or not a hypothesis is supported. Nothing in science is proven, as discussed in the introductory chapter, but hypotheses are supported or rejected after a mathematical analysis of the data. Experiments are set up using a *null hypothesis*, which is the opposite (or absence of relationship) of the *real hypothesis*. A null hypothesis contends that there is no effect or no change due to a potential treatment. If the real hypothesis asserts that variable #1 affects variable #2, the null hypothesis would state that variable #1 does not affect variable #2. If the null hypothesis is not supported then the real hypothesis can be accepted. Again, the real hypothesis is not proven, only accepted insofar as the data permits at this point in time.

There are two types of errors in interpreting experimental results and rejecting hypotheses. *Type I error* incorrectly rejects a hypothesis that is actually correct. Type I error would deprive people of a rheumatoid arthritis drug unnecessarily because the real hypothesis (that it would help) would be incorrectly rejected. *Type II error* accepts a hypothesis that is actually wrong. In the case of a drug's effects on rheumatoid arthritis, Type II error would accept that a drug helps when it actually does not. One can argue which error is worse, but it depends on the scenario.

Statistically, a *significance level* can be assigned to each study or test by the scientific community. The significance level is defined as a level of error that is acceptable. It is written in the form of a decimal number and gives the percentage chance that the results are in error. For example, a significance level of 0.05 is equal to a 5% chance that the results are in error.

Medical tests all have significance levels attached to their findings and prognosis. I recall my father's *prostate-specific antigen (PSA) test*, which looks at antigens (proteins) for cancer of the prostate in the blood. It has a significance level of 0.2, which means that it is acceptable that its results are wrong 20% of the time. I was upset with his test because, even though it showed OK, we were only 80% certain that he was not harboring a cancer. That said, error is inherent within any scientific test.

Each study, based on the limits of science and on the severity of consequences for the chances of error, assigns a significance level. In science, there are many approaches to limit error. For example, in the case of my father's prostate, the M.D. physically checks the prostate for shape and size. Generally, smoother indicates less cancer than rough. This "digital" exam triangulates with the PSA and symptoms to further reduce chances of error.

## Measures of Central Tendency

In looking at data, the ways in which the results are distributed show patterns and, of course, the effects of the independent variable on the dependent variable(s). There are several ways to represent the data.

The first is by looking at a single number to get an idea about the overall pattern. There are three numbers that give such an idea: the mean, the median, and the mode. The *mean*, or average, is defined as: $\Sigma x/n$, which is the sum of all the numbers divided by the number of measurements. The mean is the most important number in statistical tests. It is used in determining the significance of the results. The *median* is the middle number in a series of numbers. The *mode* is the most frequently occurring number in a series of numbers. Both the median and mode are rarely used statistically. The median, however, can be more representative of a larger sample than any *central tendency* measure. For example, in calculating average salary, a small group of rich people can make billions while most are poor. The average salary will still be high. Often, average household income is reported and this tells *little* about the overall population's earnings. Instead, the median would tell how people are really doing in the "middle" income bracket. Yet, for the most part, means are calculated to determine scientific results.

## Analysis of Distributions

What the measures of central tendency do not show is the *distribution*, or dispersion of numbers around the mean. Simply stated, the more the scatter in the data around the mean, the weaker the pattern and thus the weaker the effect of the independent variable. A narrow distribution shows a stronger effect of the independent variable. Please see the data dispersion graphs that follow in **Figure 4.1**.

Which graph shows a stronger effect by the independent variable: "X"? Yes, graph A is narrower, so the results are more streamlined around the mean. The *variance* and *standard deviation* are both measures of dispersion. The variance is calculated

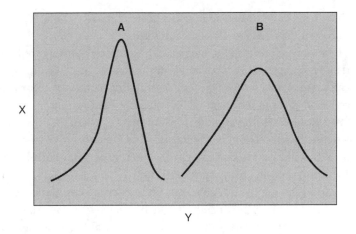

**Figure 4.1**   Distribution Graphs

by subtracting the mean from each score (x), adding them all together and squaring these differences then dividing by the total number of measurements (n) minus one. Simply dividing by the number of measurements tends to underestimate the variance.

$$\text{Variance} = \frac{\sum(x - \bar{x})^2}{n - 1}$$

Standard deviation is calculated by taking the square root of the variance. The reason for using the standard deviation in statistics, and not the variance, is because it retains units of measure (height in centimeters, days of rainfall, etc.). Calculating the square root compensates for squaring the measurements in the variance calculations. The formula for calculating standard deviation is shown below.

$$\textit{Standard Deviation} = \sqrt{\frac{\sum(x - \bar{x})^2}{n - 1}}$$

## Statistical Tests I: The Correlation

Many scientific investigations utilize the *correlation*, which is defined as a simple relationship between two variables. Variable "x" is related to variable "y" is a correlational statement. Correlations are used ubiquitously throughout science areas. For example, the area of *behavioral genetics* (a study of the genetic components of behavior) regularly utilizes correlational analyses to compare the effects of the environment vs. genetics on various behaviors. Consider the contribution of genetics to human IQ scores. Correlational analyses indicate that over 50% of IQ is genetically based. How do scientists make such a determination?

Through looking at groups of *monozygotic* (identical) *twins*, reared together vs. reared apart and comparing IQ scores between the twins of each group, a correlation coefficient can be calculated for each group. A *correlation coefficient*, represented by the letter *r* is a common index of the linear relationship between two variables. It expresses the ratio of two variables with variance in common (covariance), to variation of the two variables considered separately. Correlation coefficients range in value between $-1.0$ to 0 to $+1.0$, with the negative values representing negative correlations and the positive values representing positive correlations.

See the *scatterplot* graphs in **Figure 4.2**. The closer the correlation is to positive or negative 1.0, the stronger the linear relationship between two variables. Graph A shows a strong 1:1 positive relationship between the variables. This means that both variables increase together and conversely decrease together. The slope of a perfect relationship is a 45 degree angle, as shown in graph A. Graph C shows an equally strong negative relationship between the variables. This means that as one variable increases the other decreases. Graph B shows a correlation coefficient of 0, which means there is no relationship between the two variables being measured. Graph D depicts a weaker

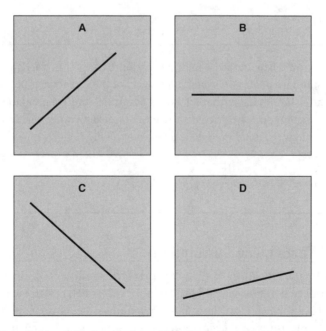

**Figure 4.2** "Best-fit Lines" for Scatterplots of Randomly Positive, Negative and Zero Relationships

positive relationship, indicating that one variable is related to the other to a smaller extent than 1.0. The slope of graph D is less than a 45 degree angle.

Consider the correlational values between siblings' IQ scores. Identical twins reared together have an $r$ value of 0.87 while identical twins reared apart have an $r$ value of 0.72.[3] The strength of a linear correlation ($r^2$) can be estimated by simply squaring the correlation coefficient ($r$). The result shows the proportion of $y$-variability that is associated with the $x$-variable. The $r$-squared value is equivalent to the percentage that one variable is related to the other. That is, the strength of the linear association between $x$ and $y = r^2$. Based on the sibling IQ score correlations, we find that for:

- identical twins reared together $r^2 = (0.87)^2 = 0.76$
- identical twins reared apart $r^2 = (0.72)^2 = 0.52$

This means that 76% of the variability in IQ scores among identical twins reared together is accounted for or explained by both the genetics and environment of the twins. Conversely, 52% of the variability in IQ scores among identical twins reared apart is accounted for or explained by genetics. Therefore, when subtracting, $76 - 52 = 24\%$ ([genetics + environment] − genetics) of IQ is due to the environment. If 52% is genetic, then the remaining 24% of influence on IQ scores falls outside of both genetics and environment. Concisely stated, the $r^2$ value represents the percentage chance that one variable is due to the other. Thus, IQ is dependent on one's genes 52% of the time, according to the stated findings.

The measure of the strength of independent variable effects in an experiment is estimated with a statistic, ***estimated omega squared***, which expresses the variation attributed to the treatment manipulation (or manipulation of the independent variable) as a proportion of the total variation in the experiment.[4] Larger proportions indicate stronger independent variable effects observed in the experiment. The primary reason for using this measure is so that research studies conducted with different sample sizes can be directly compared, as it is not affected by sample size. The relationship of these measures of independent variable magnitude is represented in the **Figure 4.3**, showing the degree to which one variable explains or predicts the variability in another variable, with two overlapping circles.[5]

We can express the degree to which $x$ and $y$ are not associated, or deviation from a linear correlation by calculating a value called ***residual variation***.[6] This quantity is obtained by subtracting $r^2$ from 1:

---

residual variation $= 1 - r^2$

---

Using our correlation between sibling IQ scores, the residual variation in IQ for twins reared together is $1 - (0.87)^2 = 1 - 0.76 = 0.24$ and for twins reared apart, $1 - (0.72)^2 = 1 - 0.52 = 0.48$. This quantity shows that 24% of the variability in IQ among the

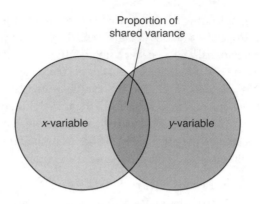

**Figure 4.3** Venn Diagram of Proportion of Shared Variance

first group is not explained by genetics or a similar home environment.[7] In the second group, 48% of the variability in IQ is not explained by simple genetics.

In interpreting the correlational results for the overall contribution of genetics or environment to IQ and thus possibly intelligence of humans, the data are quite interesting. The squared correlations show that a little over half (52%) of our intelligence is genetic in nature. When subtracting from the coefficient of the twins reared together (76%), it appears that a further 24% of intelligence is due to the home environment of children. But what is intelligence? How important is IQ?.

Correlational studies are fascinating because they shed light on patterns or relationships occurring in the natural world. Of course, a correlation of zero is also important to support a null hypothesis or eliminate relationships that are not important to a research question. Consider the fact that unrelated individuals' IQ scores have a correlation coefficient equal to zero. This is an important baseline for comparing the strength of $r$ values. Non-twin biological siblings reared apart have a weaker $r$ value of 0.22, showing the relative strength of the genetics of brothers' and sisters' relatedness and IQ.[8]

Additionally, consider a negative correlational relationship, in which one variable increases in value and the other decreases. As an example, let's contemplate a person getting mugged on the street. Psychologists have determined correlational relationships predicting the helping behavior of those witnessing a mugging based on the number of bystanders. If there are more people watching the mugging, what do you predict will be the outcome? Will it be better for the victim to have more or fewer people around to help? If you answered fewer, you were correct. As the number of individuals increase in any given area, the number of onlookers willing to help decreases to a point where often no one will help the victim.[9] This relationship is shown in the following graph in **Figure 4.4**.

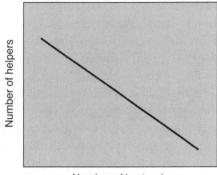

**Figure 4.4**    Helping a Mugging Victim

## Deception and Correlation

Consider a person in the front of your classroom who raises her hand and instructs you to raise your hand as well when she does. To the onlooker it would appear that you raise your hand whenever the person in front of the room does. It would appear that she is *causing* you to raise your hand. This is the essence of the problem with correlational research. Relationships do not mean that one is causing the other to occur. What if you were to find out that the person in front of the room is actually only raising her hand because someone outside the window (whom you cannot see) is doing so? In actuality it was *that* person who was really controlling your behavior. While it appeared that you were under the control of the person in front of the room, you were actually controlled by a person outside of the window. This is a demonstration I perform for my class to show the weakness of correlations. The classroom and its occupants represent the scientific community—that is, that group which observes and assesses the study from their point of view.

While correlation shows patterns, *it does not imply causation*. The actual cause was the variable outside the box—the person outside the window. So too, in science, variables unseen by simple observation may be the cause of the relationships under study. Correlations do not explain cause and effect, and other variables may intervene. Consider the correlational statement: Women who drive Mercedes cars are less likely to die from breast cancer. Obviously the Mercedes-Benz does not fight breast cancer. The car cannot physically have any impact on the disease course. Upon further analysis, we see that women who drive expensive cars also have a higher socioeconomic status and thus greater access to better health care. In this way, their survival rates from cancers are better. The correlation can thus be deceptive.

Correlational studies are used extensively in medical research because linear correlations are the easiest statistic to use when looking at large numbers of subjects. Actual experiments with more complex tests are more expensive and often unethical. As a result, medical research is often relegated to correlational reviews.

"Almonds are good for HDL (good) cholesterol numbers," "Bypasses are better for longer-term survival of cardiac patients than angioplasties," and "Atmospheric mercury deposition tends to be higher downwind from large industrial areas"[10] announce findings from recent studies showing interesting trends based on correlational research. Such pronouncements are popular and give instant assumptions for the media to fixate upon. However, there are dangers in drawing conclusions from correlations. The above studies may be scientifically sound but they show how correlations can be used to make claims that only appear quite certain in their conclusions.

## Scalar Transformations

*Scalar transformations* are changes made to scales on a graph or a set of data meant to manipulate the appearance of the results. They are another method used in the analysis and presentation of data that can consciously or unconsciously misrepresent the findings of an investigation. Consider the graphs in **Figure 4.5**. The first graph shows little relationship or change in the variable as a result of the other. The second, altering the scale of the *y*-axis, exaggerates the slope of the line and emphasizes variation in the data represented on the *y*-axis.

## Interpolation vs. Extrapolation

Another risky tendency for researchers and the public is to make certain assumptions about data from scientific investigations. *Interpolation* refers to the process of filling in, or making generalizations, between two data points in a table or graph.[11] Whenever a

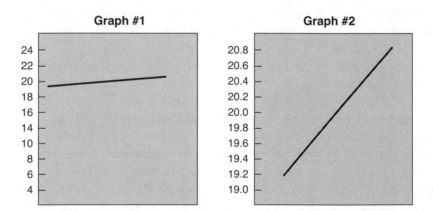

**Figure 4.5**   Scalar Transformations Between Two Graphs

| Table 4.1 | Helping a Mugging Victim |
|---|---|
| Number of Onlookers | Actual Helpers in Mugging |
| 10 | 5 |
| 20 | 4 |
| 30 | 3 |
| 40 | 1 |

line on a graph is drawn, an interpolation is made. In the example of helping behavior cited earlier in the chapter, perhaps a data table would look like **Table 4.1**.

An interpolation would surmise that there would be two helpers for between 30 and 40 onlookers. Interpolation leads to the assumption that this is the case. Perhaps, in practice, there would be a larger number of helpers. We really don't know, but based on interpolation of the available data one expects two helpers to aid the mugging victim.

On the other hand, *extrapolation* is predicting beyond the data on the graph. It is based on trends revealed by the data but makes predictions beyond the limits of the data.[12] Extrapolation is risky because the study does not provide data that can support such claims. It can be argued, however, that a study without some generalization or attempts to make a prediction based on the data is often uninteresting and not useful. Extrapolation leads to future tests to see if the prediction holds true. Consider the helping behavior example. If we extend the table downward and predict that zero helpers would come to the victim's aid, it has applicability to a common situation (if mugged on a busy New York City sidewalk, don't expect any help). Further studies on helping are likely to be stimulated. Extrapolation is, however, untested and so it is without mathematical support.

## Statistical Tests II: Comparing Group Means

In the early years of statistical design, researchers used the *t-test* to analyze the results of two-group studies.[13] The t-test compares the means of two groups to detect a difference between them, statistically. It has only the power of a correlation. Thus, a more powerful statistical test, called the *Analysis of Variance (ANOVA)*, was developed. The ANOVA is a test that identifies and isolates the sources of variation during hypothesis testing.[13] Earlier in this chapter, the weakness of correlations in not controlling for extraneous variables was discussed. The ANOVA seeks to reduce extraneous variable effects by comparing the means of three or more groups in an experimental design. Very briefly, it is a process of isolating two sources of variation within data: one source showing the effects of the independent variable and the unavoidable effects of experimental error, and the other source reflecting the effects of experimental error alone. By subtracting the two numbers the effects of the independent variable are determined. Determination of the effects of the independent variable is the point of any experiment and data analysis.

The ANOVA's statistical methods use means and standard deviations discussed earlier in the chapter. The results of the test give what is termed an *F-ratio*. The F-ratio is equal to the sum of the variation between the groups divided by the sum of the variation within the group.

---

F-ratio = *Sum of Squares Between Groups/Sum of Squares Within Groups*

---

More variation within the group produces weaker results. "Within group" variation is the amount of difference between individuals placed within a treatment group in an experiment. That difference is not due to the independent variable but to "other" variables that are inherent within the subjects. The statistical details of the ANOVA and its affiliated tests are beyond the scope of this text. However, its importance in showing how researchers statistically isolate as many independent variables as is possible is vital to establishing valid scientific results.[13]

Human research presents a particularly more difficult isolation of variables. It results in much a higher amount of within group variation. In the accompanying pie chart in **Figure 4.6**, consider the bacterial genus *Streptococcus* used in microbiological research. The variation between individual bacterial organisms is much smaller as compared with humans. In the figure, pie #1 represents variations in the bacterial experiment and pie #2 shows variation in the human populations.

Differences between individual bacterial organisms are limited because they can be taken from the same colony (are therefore genetically similar) and are sometimes even clones of one another (eliminating differences between organisms).

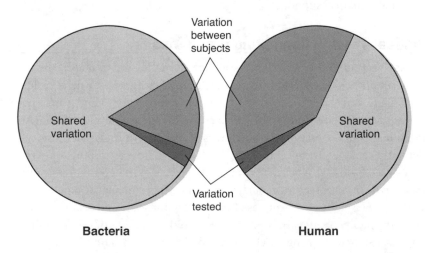

**Figure 4.6**   Bacteria vs. Human Variation

Minimizing the individual differences between subjects helps isolate the effects of the independent variable. Humans, being more complex creatures, have more "within group" variation. This decreases the F-ratio of the ANOVA and thus the strength of the results. If differences, for example, between groups of humans eating almonds vs. no almonds exist in their high-density lipoprotein (HDL) cholesterol levels, there are many variables that could have intervened. One group could have cheated and reported fewer almonds eaten. Humans are difficult to study because of the many differences between individuals often unaccounted for in the design. Thus, disciplines studying human trends are often termed "soft" or social sciences. Psychology, anthropology, sociology, and economics are exemplary. They are not weaker, of less importance, or of less rigor but do lack the statistical power behind their results and conclusions.

# ■ KEY TERMS

| | |
|---|---|
| angioplasty | monozygotic twins |
| Analysis of Variance (ANOVA) test | null hypothesis |
| behavioral genetics | prostate-specific antigen (PSA) test |
| central tendency | real hypothesis |
| correlation | residual variation |
| correlation coefficient | scalar transformations |
| distribution | scatterplots |
| estimated omega squared | significance level |
| extrapolation | standard deviation |
| F-ratio | sum of squares between groups |
| interpolation | sum of squares within groups |
| mean | type I error |
| median | type II error |
| mode | variance |

# ■ PROBLEMS

1. Compare and contrast the following sets of terms (give one way the terms are the same and one way the terms are different):
   a. Variance and Standard Deviation
   b. ANOVA and t-test
   c. Mean, median, and mode
   d. Interpolation and Extrapolation
   e. Type I and Type II error
   f. Real and Null Hypothesis

2. "Correlation does not imply causation." Discuss this quote.
3. Which *r* value represents a stronger correlation?
   a. −0.8
   b. 0
   c. +0.6
   d. +1.6

## ■ REFERENCES

1. Sherman, D. and Salisbury, J. 2011. *The west in the world* (p. 455). New York: McGraw-Hill.
2. Staff and wire reports. *USA Today*. Sept. 5, 2001: 7D.
3. Morris, C. and Maisto, A. 2001. *Understanding psychology*, 5th ed. (pp. 254–257). Upper Saddle River, NJ: Prentice Hall Publishers.
4. Keppel, G., Saufley, W., and Tokunaga, H. 1992. *Introduction to design and analysis: A student's handbook*, 2nd ed. (pp. 180–183). New York: W.H. Freeman and Company.
5. Ibid, p. 481.
6. Ibid, p. 483.
7. Morris, C. and Maisto, A., 2001, *Understanding psychology*, 5th ed. (p. 255). Upper Saddle River, NJ: Prentice Hall Publishers.
8. Ibid, p. 256.
9. Ibid, pp. 511–513
10. Kroenke, A. 2003. *Atmospheric mercury deposition to sediments of New Jersey and southern New York State: Interpretations from dated sediment cores*. Ph.D. Thesis. Troy, NY: Rensselaer Polytechnic Institute.
11. Allen, G. and Baker, J. 2000. *Biology: Scientific processes and social issues*, Hoboken, NJ: John Wiley & Sons.
12. Ibid.
13. Keppel, G., Saufley, W., and Tokunaga, H. 1992. *Introduction to design and analysis: A student's handbook,* 2nd ed. New York: W.H. Freeman and Company.

# CHAPTER 5

# The History of Science

"We brought the boy in an hour ago. He was bitten by a mad dog!" exclaimed the nurse. It was the last time Dr. Pasteur could stand to see a patient die from an animal bite. Dr. Louis Pasteur, in his 19th century hospital, knew the symptoms of rabies: fever, headache, tiredness, drooling. He was angrier this time. "I cannot let another child die!" he thought. Asking his colleagues, Dr. Vulpain and Dr. Grancher, "What do you recommend I do for the boy?" Both responded, "Test the vaccine. Otherwise, he'll die; there is nothing to lose." "But is it ethical?" Pasteur wondered. His previously conducted experiments on dogs seemed to imply that it could work. In the laboratory, he discovered that if the spinal cord of a dead rabid dog was preserved it lost its infective power after about two weeks. If this spinal material was then injected into a rabid dog, the dog was able survive to the infection. "How macabre," he thought, to inject a human with ground up spinal cord from a dog. The alternative, however, was guaranteed death by rabies for his patient. Dr. Pasteur injected the boy with the spinal cord mixture. A long and watchful night passed, but at the end of it the boy was still alive. On succeeding days more injections were made. The boy slowly improved, became stronger, and on the third month Pasteur announced that the child was out of danger and was the first rabies victim to survive in human history.[1]

## Introduction

Louis Pasteur's nineteenth-century discovery of the rabies vaccine in the story above illustrates the human struggle to use the power of the mind to overcome natural and societal proscriptions. The seeds for the future of scientific thinking are born of this quality, as traced in the history of science in this chapter. It was best described in 1665, at the end of a period called the "*scientific revolution*" in Europe, by mathematician and scientist Blaise Pascal (1623–1662). Upon retirement, Pascal began writing these thoughts: "Man is but a reed, the most feeble thing in nature; but he is a thinking reed . . . All our dignity consists, then, in thought . . . by thought, I comprehend the world."[2] The quote reflects the frailty of the human condition but the strength of human thought.

Pascal's words embody the essence of the emergence of a way of thinking about the natural world. It was a movement from the power of physical strength to the power of the mind. This theme highlights the struggle to overcome nature through innovation in the history of science. The future of science lies in its past: The passion of the great scientists, their struggles against society, and their creativity in discovery all point to the characteristics needed to alter paradigms and propel scientific thought in our future.

The scientific revolution, which lasted roughly from 1540 through 1690, comprised a period of skepticism about existing knowledge of how the world worked (**Table 5.1**). It was an era of rapid advance in Europe, which questioned authority and set the foundations for modern science. During this period, systematic and organized research

| Table 5.1 | Key Dates in the Scientific Revolution |
|-----------|----------------------------------------|
| 1543 | Copernicus's heliocentric model published |
| 1543 | Vesalius's publication of *On the Fabric of the Human Body* |
| 1609–1619 | Kepler's three laws of planetary motion |
| 1633 | Galileo's trial |
| 1662 | English Royal Society founded |
| 1673 | Van Leeuwenhoek's animalcules observations |
| 1675 | Boyle's laws discovered |
| 1687 | Newton's publication of *Principia* |

in thinking about the natural world led to discoveries and methods still guiding contemporary science.

There were scientific advancements far before 1540 in a host of ancient civilizations from the invention of the alphabet and the compass to bronze and iron metallurgy. Star and planetary motion principles came from the Chinese; glass, metal, and plastics development from Islamic cultures; and the ancient Greeks engineered waste removal and discovered that the Earth was round through systematic study. In fact, by 2000 B.C. in China, they understood that the heart was a pump through which blood was circulated around the body and the *Medicine Book of the Yellow Emperor* distinguished between 28 different kinds of pulses.[3] Humans had been thinking and discovering since their development but science as a formal, organized field was unknown to them until much later in human history.

## Aristotelian Views

Until the 1500s, science did not yet exist as a discipline. Instead, people studied *natural philosophy*. Natural philosophy was the study of the universe, focusing on how it functions and its purpose. Natural philosophy was the precursor to science. The views of *Aristotle* (384–322 B.C.), the great Greek philosopher of the fourth century B.C., formed the major tenets of natural philosophy. Medieval theologians, such as *Thomas Aquinas* (1225–1274), integrated the ancient views of Aristotle into harmony with Christian ideology. Aristotle and Aquinas emphasized that the Earth was the center of the universe and that humans were special creations in that center. A view of the universe showing the Earth as its center is termed the *geocentric model*, as shown in **Figure 5.1**.

Aristotle's observations explained the movements of the moon, the sun, the five known planets, and the stars around the sky. He believed that they all rotated around a fixed Earth center. Beyond this was Heaven, the throne of God and the saved souls. In this view, Earth was impure, with its uneven surfaces. In contrast, the heavenly bodies circling the Earth were the example of God's perfection, smooth and pure.[4] This is where Heaven was thought to be.

Aristotle's views made sense because they could explain how the observed universe, for the most part, behaved. Lower levels of argumentation, observation and classification, were the stages of thought forming Aristotelian natural philosophy. This is not to be taken pejoratively. A look at the sky confirmed that the sun and the stars moved around the Earth in certain paths in the sky. The universe was understood. It was clear and simple and humans on Earth were at the center of God's concern.

Aristotle felt that ideas could not exist outside of physical manifestations. In other words, if something was not sensed, it should not be studied. To study anything required an actual entity and he divided knowledge into categories which remained the guiding principle for learning for about 2,000 years: *ethics* (principles of social life), *metaphysics* (laws of the universe), and *natural history* (nature).[5] Aristotle used logic based on observation to classify the natural world, categorizing life's creatures

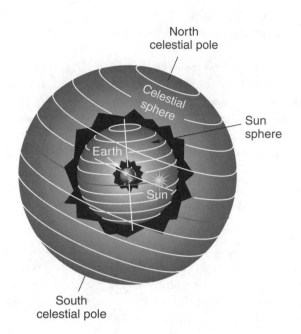

North
celestial pole

Celestial
sphere

Sun
sphere

Earth

Sun

South
celestial pole

**Figure 5.1**   Geocentric Model: The Greek model of the universe shows the sun as a sphere
that moves around the stationary Earth inside the celestial sphere of stars. The axis of
the sun's sphere is tilted with respect to the axis of the celestial sphere to account for the
apparant pathway of the sun and the sky.

according to similarities and differences. He is called the "father of science" for first
basing thought on observable patterns and comparisons.

While whatever was seen was truth in the ancient and medieval eras, there were
some observations difficult to explain in Aristotle's time. For example, he could not
explain why some planets moved backwards. A second-century Greek scholar, *Ptolemy*
of Alexandria (ca. 90–ca. 168), offered a solution. His picture is shown in **Figure 5.2**.

Ptolemy argued that planets moved in smaller circles, each of which turned along
a larger circle, to create a backwards-appearing effect. This described the backwards
variant and was actually an accurate model for planetary motion, as shown in
**Figure 5.3**. Ptolemy preserved the geocentric model of the universe.

Until the sixteenth century, European scholars mostly accepted Aristotle's views of
physical nature. It is surprising that his teachings dominated almost 2,000 years of scientific
thought. But it was a very conservative intellectual community. The ideas of Aristotle and
Ptolemy were passed on from generation to generation of scholars. The ancients through
the Byzantine Empire and onto the modern medieval world held tightly to Aristotle's beliefs.
Arab scholars and even Chinese thinking about the world were in tandem with an Earth-
centered universe. It made sense. Accordingly, investigations about the physical universe
consisted only of making deductions from such long-held traditions.[6]

**Figure 5.2**   Ptolemy

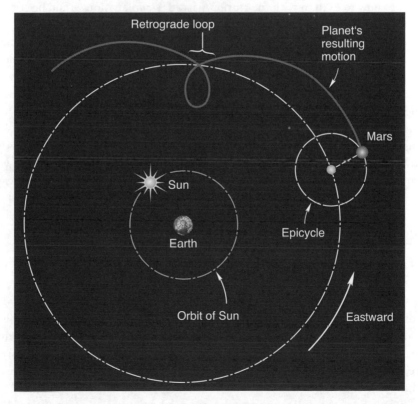

**Figure 5.3**   Mars' motion on its epicycle results in a looping path, which- when seen from Earth- appears as retrograde motion.

## Ancient Disagreement Discovered

New ideas were then uncovered to undermine this traditional view. During the Middle Ages, a search for ancient Greek writings led scholars in Europe to discover Greek authorities that contradicted Aristotle.[7] One such authority was *Plato* (429–347 B.C.), another ancient Greek philosopher. The views of Plato went against Aristotle. *Neoplatonism*, based on the ideas of Plato, emphasized that skepticism and a rejection of appearance is the way to scientific truth.

In fact, Plato was Aristotle's teacher, whom Aristotle had studied under and for whom Aristotle had great respect. While Aristotle valued Plato's views, Aristotle felt that only the observable should be under consideration. Plato emphasized abstract postulates that led to disagreement. Aristotle disagreed with Plato. Aristotle's views of the world avoided the abstract and looked only to the tangible. To show the contrast, consider the study of leaves on trees in a forest. Aristotle would classify the leaves according to shapes and give them names. Plato would argue that the shapes mean something, perhaps related to their historical development, to consider the larger picture of how trees fit in with the world. Aristotle's explanation would be concrete whereas Plato would play with possibilities.

Plato was a student of *Socrates* (470-399 B.C.), an ancient Greek philosopher as well, and arguably, the founder of the modern scientific method. Socrates developed his ideas as a reaction against the "groupthink" of the times. In Greece, democracy was actually quite dictatorial and tolerated little dissent. Socrates was very outspoken and would roam the streets questioning people and their ideas. He had a following and many listened to him. In questioning his followers and students, he would refute their answers and use argumentation to help bring them to a greater understanding of truth. This type of questioning and answering is now called the *Socratic method*. Socrates never wrote down his ideas but Plato recorded his many dialogues to preserve his teacher's way of thinking. The following shows the relationship of the three natural philosophers:

---

Socrates → Plato → Aristotle.

---

Of course, Socrates enraged the powers of the ancient Greek democracy. They were threatened by his questioning of a variety of issues at the time, including local wars and economic decisions. Socrates was tried and found guilty of corrupting young minds. He was sentenced to death by drinking a cup of the deadly poison hemlock. Socrates had the opportunity to flee but to this he replied, ". . . Life without enquiry is not worth living. . ."[8] Socrates is best remembered in the scientific community for these words and it has sparked, in science education, a modern movement toward teaching and assessing inquiry.

Discovery of the philosophers who opposed Aristotelian science gave birth to the scientific revolution. It was a revolution because it was a complete change in thinking. Plato and Socrates advocated questioning of what was known. Plato's emphasis of moving beyond mere appearances to obtain true knowledge is based, in part, on **Hermetic doctrine**. Hermetic doctrine is based on the writings of **Hermes Trismegistus,** an ancient Egyptian priest. It states that all matter contains a divine spirit. That divine spirit or energy needs to be studied in order to be unleashed to give knowledge. Therefore, knowledge exists in nature and through studying the natural world, people learn.

The practices of Hermetic doctrine moved beyond the obvious and sought to find God in nature and in mathematics. It required an inquisitiveness that moved beyond the observable to what was inside the natural world. Neoplatonic-Hermetic thought thus encouraged a number of investigators to search for truth in a variety of areas. Uncovering what was in nature's substances was the source of the new information. *Alchemy* (study of transforming metals into gold; early chemistry), *astrology* (study of how stars affect humans; early astronomy), and magic (manipulation of human perception) witnessed a boom into new ways of questioning. This type of thinking started people tinkering with ideas. That was the start of modern science.

## Changing Viewpoints

In the sixteenth century, the age of discovery led to communication with other societies. Exploration of other regions overseas exposed Europe to other value systems. Thus, Europe changed with incoming products and ideas. The invention of the printing press, dissemination of books, and contact with other cultures through colonization led to interchanges that built upon existing ideas. For example, the discovery of the new world disproved Ptolemaic geography. New navigation methods were sought for overseas exploration. It stimulated demand for new instrumentation, such as telescopes, and encouraged research in mathematics and astronomy to improve travel. Internally in Europe, the growth of medieval universities formed a society of philosopher-scientists to challenge the larger communities. In these ways, western philosophy grew to change the assumptions of Aristotelian science.

Systematic data collection and a search for newer explanations prevailed in Europe. At the same time, by the sixteenth century, non-Western societies in Islamic, Chinese, Japanese, and Indian areas no longer questioned their traditional ways of thinking. Thus, they fell behind in scientific advancement in the 1600s, allowing Europe to take the lead.

### The Copernican Revolution
On June 22, 1633, Italian scientist *Galileo Galilei* (1564–1642), shown in **Figure 5.4**, was on trial for popularizing the view that the sun, and not the Earth, was the center of planetary motion. He was sentenced to house arrest in Florence for the rest of his life and forced to reject his "false opinion that the sun is the center of the universe and

**Figure 5.4** Galileo Galili

immovable, and that the Earth is not the center of the same."[9] The scientific revolution was born out of struggles parallel to those Galileo experienced in this trial. But the scientific revolution was not an incident but a metamorphosis of thought. Western challenges to the assumptions of the times led to much controversy.

In the 1500s, astronomy and physics attracted the most attention from scholars because it linked with the heavens. At this time, Aristotle's tenets no longer were able to fully explain new observations and planetary calculations. As English poet John Donne complained in a poem in 1611, "New philosophy calls all in doubt." This quote shows a frustration between the larger society and the great thinkers of the scientific revolution.

In line with the new movement, Polish clergyman *Nicolaus Copernicus* (1473–1543) published, in his last year of life, a mathematical formulation that the Earth was not at the center of the universe. Instead, he believed that the sun was at the center and that "what appears to be the motion of the sun is in truth a motion of the Earth."[10] His portrait is shown in **Figure 5.5**.

Copernicus's work is termed the *heliocentric model* of the universe because it places the sun at the center. It implicitly deemphasizes the importance of the Earth and was denounced by religious and government leaders in Europe in power at the time. Society did not accept Copernicus's book, *De Revolutionibus*, or anyone who would defend it. The accepted geocentric model of the universe placed Earth and humans in the center. To question this would be to question the important link between God and humans. His original model is shown in **Figure 5.6**.

Some supporters of the heliocentric model, however, worked despite public disagreement and often punitive action. *Giordano Bruno* (1548–1600), an Italian monk, extended Copernicus's ideas and publicly taught that the universe was infinite

**Figure 5.5**  Copernicus

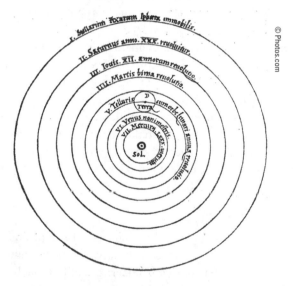

**Figure 5.6**  The sun-centered model of the universe that appeared in Copernicus' De Revolutionibus. It is said that a copy of the printed book was rushed to Copernicus so that he could see it as he lay on his deathbed

and had no surfaces and no end. He was burned at the stake but his ideas led to more modern thinking and our current view of the infinite size of our universe.

## A New Physics

Later in the revolution, *Johannes Kepler* (1571–1630), a German Hermetic scholar, used mathematics to understand the universe. His studies led him to the three laws of planetary motion that are still utilized today. First, that planets move in ellipses around the sun; second, that a planet's speed varies with its distance from the sun; and third, that the relationship between moving planets can be expressed mathematically. Kepler's thinking led to a view that the universe could be explained mathematically and that there is an order in the way the physical world operates. Kepler and Galileo communicated with each other through letters. They had a common interest seeing the mathematics in their observations. Kepler encouraged Galileo to use this way of thinking to study and publish his findings. They were both frightened for their safety. Kepler's portrait and instrumentation are shown in **Figure 5.7**.

The newly developed telescope was a useful tool for Galileo. With it, he discovered that the moon's surface was in fact not smooth but much like Earth's, imperfect with crater holes and ridges. This juxtaposed an Aristotelian view that heavenly bodies were perfect. Galileo also revealed that there were moons around Jupiter and that sun spots flared in different places on the solar surface. This brought into question the perfection of matter outside of Earth. It also showed that there was much more to learn about the heavenly bodies. The universe, however, became less perfect, and Earth, less important. Religious authorities at the time were enraged.

Through experimentation and mathematical analysis of the data, Galileo rolled balls down inclines and studied how motion could be described mathematically. While Galileo worked within religious groups of the times, making concessions on points to get his book *Dialogue of the Two Chief Systems of the World* published, the Roman inquisition still ultimately forced him to renounce his views. This tension between the larger sociological entities and the smaller, science community is a theme of this text. Galileo's suffering did scare scientists but they still communicated to advance his ideas.[11] A timeline showing the changes in scientific thinking about the structure of the universe is given in **Figure 5.8**.

## The Birth of Newtonian Physics

*Isaac Newton* (1642–1727), a devoutly religious natural philosopher, studied at Cambridge University in England. He saw the advantages of scientific methodology but wanted to harmonize Christianity with modern scientific thought. His contributions are stunning. All of modern mechanical physics is still based on Newton's observations and deductions. **Figure 5.9** shows Newton and his work with light and prisms.

Newton developed *calculus*, a mathematical area of knowledge using special symbolic notations, with a focus on studying rates of change. He formed the laws of motion, which define Newtonian physics: inertia, acceleration, and action/reaction. These will be discussed in more detail in other chapters. His work relating the attraction between two bodies based on their mass and distances was published in 1687, in *Principa,* his defining work. Engineering projects, acceleration in car testing, and studying attraction of planetary bodies all use Newton's ideas today. The first part of any physics course is termed "*Newtonian physics*" because all motion is based on his principles.

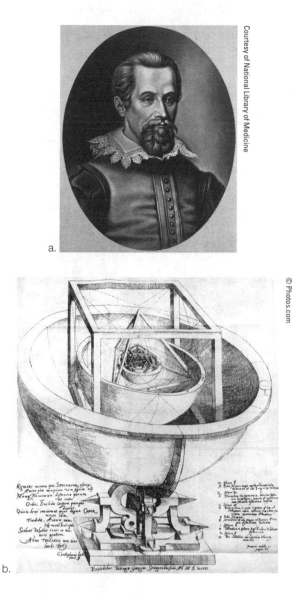

a.

b.

**Figure 5.7** Johannes Kepler and his Instrumentation

In particular, Newton's unifying theory of motion was accepted and built upon through modern times. He emphasized that the universe consists of only two things: matter and motion. Until Albert Einstein, who modified such thought to include space and time as variables, Newtonian physics was the only accepted way to understanding the universe.

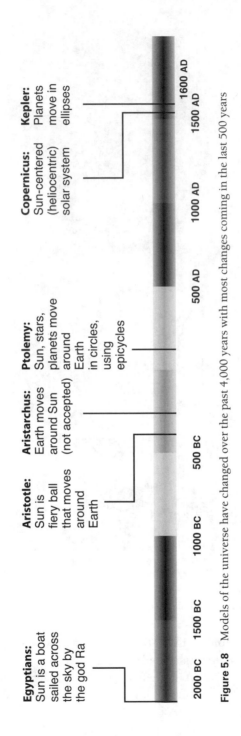

**Figure 5.8**   Models of the universe have changed over the past 4,000 years with most changes coming in the last 500 years

© Photos.com

**Figure 5.9**    Sir Isaac Newton worked in many fields. In this painting, he is shown experimenting with a prism to investigate the nature of light

## Revolution Spreads

Other disciplines changed their ways of studying phenomena. Medicine, anatomy, and chemistry saw an implementation of Plato's skepticism. Natural philosophers investigated these areas in unorthodox ways. To illustrate, *Paracelsus* (1493–1541), a Swiss alchemist-physician, rejected the Aristotelian idea that water, fire, earth, and air made up all matter. Instead, he looked at specific chemical imbalances that caused diseases and administered arsenic and mercury to cure various ills. While his medical treatments were not close to being successful, he laid the foundation for modern pharmacology. His work showed that chemicals make up the body in different proportions. Thus, chemicals can be given to cure disease, he surmised. Paracelsus's contributions encouraged further research and experimentation to investigate disease control through drug treatment. This was therefore also the birth of modern medical drug research.

Other investigators founded the study of anatomy. *Andreas Vesalius* (1514–1564) wrote a 1543 treatise, "On the Fabric of the Human Body," which used observations and assumptions from his human dissections. Stage by stage, the work detailed aspects of human anatomy. He would secretly pick up bodies on the roadside after they had been hanged or even dug up human corpses. Vesalius endured a great deal of conflict with contemporary scholars and physicians because of his obvious human dissection. His detail was too clear. This was unacceptable at the time. Society saw the human body as sacred, not to be studied or manipulated. The social pressure caused Vesalius to give up on his research. Later, William Harvey (1578–1657), an Englishman, extended the work on cadavers, observing that the human heart was the pump of the body and that the valves in the heart served to prevent back flowing of blood. Harvey studied dogs, pigs, lobsters, and shrimp, making comparative observations. He found patterns, mostly anatomically, between and among the organisms he studied. His work gave birth to the study of modern *comparative anatomy*.[12]

On a smaller scale, *Anton van Leeuwenhoek* (1632–1723), a Dutchman, observed a variety of specimens under the newly discovered microscope. In the 1670s, he described the "little animals or animalcules" in water samples taken from lakes. These are what we now know to be microorganisms or microbes. He saw bacteria and fungi on cheese, bacteria in his saliva, and sperm from his own samples. He described the objects he saw as "little eels or worms, lying all huddled up together and wriggling . . . This was for me, among all the marvel that I have discovered in nature, the most marvelous of all."[13]

The discovery of microorganisms was a door to an explosion of diversity. That so many creatures lived in a world that we did not see with the naked eye was heretofore unimaginable. Complexity and diversity of cells led to a drive to find out about a whole world of "little things" and thus the birth of the study of organisms too small to be seen with the naked eye, *microbiology*. Unfortunately for science history, smaller substructures within the cell were not to be observed until the discovery of the electron microscope in the 1930s. This allowed science to see the smaller parts of the cell called *organelles*. These include structures such as the mitochondria, endoplasmic reticulum, and golgi bodies that serve vital purposes within the cell. The birth of microbiology is perhaps the largest jump to viewing the complexity that underlies our existence.

On an even smaller scale, an Irish nobleman, *Robert Boyle* (1627–1691), was interested in the smaller chemicals found in the human body. He thus laid the foundations of modern chemistry. Extending the work of Paracelsus, Boyle relied on experimentation and new instruments to begin a systematic approach to search for the basic elements of matter. He also discovered a law describing the nature of pressures and volumes of objects as they related to each other, termed today as *Boyle's Law*. He was exact in his process and his procedures, thereby setting the standards for many modern chemistry laboratory techniques.

The great thinkers of the scientific revolution set the stage for later scientific advances. It was a time of revolution because it upset the status quo of thought and led to permanent changes in investigation methods. Knowledge was no longer unchanging but subject to disapproval and dismissal.

## The Enlightenment

After the scientific revolution, an *Enlightenment* period (1733–1789) awakened a host of new intellectual movements. There were new applications of the scientific method. Research into areas other than natural philosophy began to be applied in nonscience areas. The scientific method was used to study societies (sociology), individuals (psychology), economic theory (capitalism, termed *laissez-faire* economics at the time), and women's rights movements.

Science branched into areas using rational, secular critical thinking to analyze problems. Enlightenment philosophers questioned all authority, from the Church to governmental structure and economic systems. The age of questioning reinvented Greek writings. Its achievements in free thinking were its very demise. Breaches in morality and excesses of the times were blamed on free thought and a conserving reaction followed. The public saw a return to religiosity and a rejection of what appeared to be scientific atheism. While the end of the era witnessed a restriction in thought, the optimism about science fomented during the Enlightenment transcended the next centuries. It led the way to the breakthroughs in modern scientific thought that would have been impossible without the freedom of thinking developed during this period.

## Bridging to the Twenty-First Century

The centuries after the scientific revolution still included a great deal of classification type science. *Taxonomy*, the science of classifying based on observable patterns among organisms, continued to search for order and naming creatures in nature. Much of eighteenth- and nineteenth-century science encompassed classification schemes in many areas such as: gross anatomy, plant biology, astronomy, and chemical identification. Even Darwin's theory of *evolution* in the late 1800s was based on patterned observation of organismal characteristics. Charles Darwin's evolutionary theory did not use experimental or chemical bases for his theories either. The eighteenth and nineteenth centuries started to see a shift though, from taxonomic to smaller, more indirect means of study.

The twentieth century saw a much greater emphasis on theoretical and microscopic analyses as compared with preceding centuries. The discovery of more indirect methods of observation helped facilitate research at chemical and molecular levels. The German physicist *Max Planck* (1858–1947) led the way, arguing in 1900 that matter emitted discrete quantities of energy and that matter was not fully separate from energy. This undermined Newtonian physics, which discretely placed matter and energy as separate. In 1905, German-Jewish physicist *Albert Einstein* (1879–1955) published the *theory of relativity*, which gave the mathematical relationship between matter and energy. Relativity described matter as curving the space and time around it, as shown in **Figure 5.10.**

Einstein showed his ideas mathematically and not experimentally, emphasizing that there is a great deal of energy contained within every bit of matter. This led the drive to find the energy of the atom later in the 1940s through *atomic fission* (splitting of the atom) and the discovery and development of *radioactivity* (energy emerging from the splitting of the atom). Einstein's photograph is shown in **Figure 5.11.**

Other discoveries in equipment led to advances in various fields of science. To illustrate, x-ray diffraction used by physicists led to the discovery of molecular

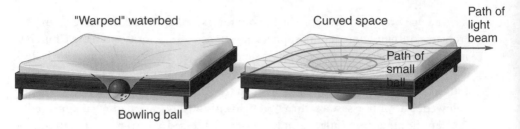

"Warped" waterbed    Curved space    Path of light beam

Path of small ball

Bowling ball

**Figure 5.10**    According to the theory of relativity, matter curves the space and time around it

© AIP Emilio Segrè Visual Archives

**Figure 5.11**    Albert Einstein

shapes in a three-dimensional form. A host of areas of research were then possible. Medical imaging of bones and other internal organs, biochemical study of enzymes and hormones, and different structures in the Earth's crust were finally being realized. Most importantly for modern genetics, new techniques could be used to explain how *DNA (deoxyribonucleic acid)*, the hereditary material of life, was able to give rise to itself as well as all of the proteins in the body. The 3-D shape of DNA (and theories of replication and protein synthesis) was discovered by Francis Watson and James Crick in the mid-twentieth century. These scientists used data to form a model of DNA. We use this model as a foundation for studying genetics today.

## Rosalind Franklin and DNA

When English scientist *Francis Crick* and American biochemist *James Watson* were working in Cambridge University in 1953, they used information from several sources to discover a new model for DNA. The brand new Ph.D., 23-year-old James Watson traveled to London and visited the lab of Maurice Wilkins at King's College. There he discovered an x-ray image of DNA taken by *Rosalind Franklin,* Wilkins's colleague (**Box 5.1**). Watson studied the image to deduce the shape of DNA to be a

## BOX 5.1    HISTORICAL NOTE: ROSALIND FRANKLIN

Why is Dr. Rosalind Franklin, a brilliant scientist, left out of so many textbooks discussing the discovery of DNA? The truth behind how DNA was discovered comes from King's College in London in 1953. Dr. Franklin was born in 1920 and earned a doctorate in physical chemistry at the age of 26, against her father's wishes. She worked at King's College, refining x-ray diffraction to produce the "Photograph 51" that helped Watson and Crick develop their model of DNA.

When Dr. Franklin was a colleague of Maurice Wilkins, she was treated as a mere helper and suffered terrible gender discrimination in essentially a male-dominated field. As such, Rosalind Franklin was never given the proper credit for her contribution to the Watson and Crick model. Although her work was published in the same issue of the journal *Nature* as Watson and Crick in April 1953, she was not given credit for her contributions to the model.

Tragically, she died of ovarian cancer in 1958, at the age of 37, 4 years before Watson and Crick received the Nobel Prize for their work. It is likely that she died for the model. She worked hundreds of hours to perfect her photographs of crystallized DNA, exposing her to large doses of radiation. The photo in Figure 5.12 shows scientist Rosalind Franklin's x-ray image of DNA. Dr. Franklin is given posthumous credit in this text.

2 nm wide helix, hence the model. Franklin's work actually gave Watson the idea for the double helix model, but she did not receive credit in the publication describing the arrangement of DNA. Unfortunately, gender discrimination leaving out credit for the work of many women in science was commonplace through much of science history. Franklin's image is shown in **Figure 5.12**.

Watson and Crick put all of the research together from the varied sources, figuring out just how DNA is inherited from generation to generation. This was the birth of

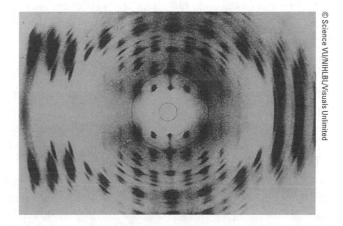

**Figure 5.12**    Discovering DNA's Helical Shape. The famous X-ray diffraction photograph of DNA by Rosalind Franklin.

molecular genetics (a combination of the fields of biology, chemistry, and genetics), the study of inheritance based on the structure and function of inherited material at the chemical level. Watson and Crick's model is used to explain many aspects of chemical inheritance.

Their model of the DNA structure explains a great deal about how DNA divides; it also explains how it affects the many activities within the cell. Their discovery is more than a simple description of a chemical: It is the basis of explaining how information is passed on from cell to cell and within cells. The model will be used through this chapter to describe how that happens.

Watson and Crick showed that DNA resembles a twisted ladder, with sugars and phosphates on the vertical parts of the ladder and certain chemicals that match up (called bases) making up the rungs of the ladder. This type of structure is known as a "double helix." DNA is a **nucleic acid**, a macromolecule that stores information—the "code of life" in strings of base sequences. The exact structure of DNA as sketched out by Watson and Crick is shown in **Figure 5.13**.

## Breakthroughs to "Smaller" Science

DNA's discovery was a breakthrough to the science of objects unseen. In particular, proteins are made by DNA and carry out all of the activities of the body. Proteins also comprise much of the body's structure: Hair, nails, enzymes to digest food, hormones to communicate internally, antibodies to defend against infections, and hemoglobin to carry oxygen and carbon dioxide in the blood are all examples of the proteins formed from the DNA. They are all three dimensional and biochemistry developed rapidly in the twentieth century to study these many substances. They are not seen easily with any instrument but are detected and studied indirectly through models and modern instrumentation. The twentieth century saw the development of applications from this new molecular science.

The developments had important implications for human health care, in particular. Through studying the many substructures in cells, newer medicines to treat, for example, hormonal imbalances and blood disorders have been developed in the past century. Twentieth- and twenty-first-century science have moved from a taxonomic focus of the early 1900s to a modern synthesis of the science areas with applications for human use.

The shift from a gross (large in size) to molecular approach to study smaller scale applications is an accelerating trend. Consider *nanotechnology,* which deals with objects on the level of $10^{-9}$ meter scale. Large amounts of data can be obtained in smaller and smaller amounts of time using such small scales. Many drug treatments for cancer, for example, could be tried in one shot using nanotechnology by placing a set of cells onto a test plate containing thousands of chips containing drugs. This allows scientists to carry

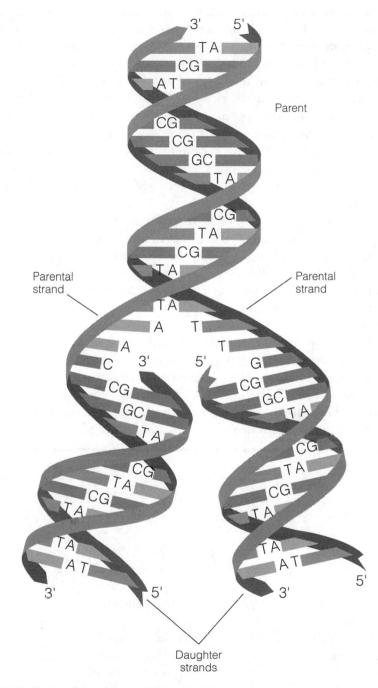

**Figure 5.13**  Deoxyribonucleic Acid Structure & Base Pairing

out experimentations that once required years of separate procedures. Even smaller than nanoscale, physicists are studying particles smaller than basic atomic matter called quarks. *Quarks* are parts of neutrons and protons (particles inside of atoms). Quarks are hypothesized to have even smaller parts, known as superstrings, which are levels that make up matter at the smallest scale. Superstrings or strings for short, are one dimensional loops of matter that vibrate as an infinitely thin rubber band. This view of matter is known as *string theory* and has been mathematically but not experimentally shown. Technologically, science is unable to "see" strings but mathematics shows that they are present in matter. Alternatively, it is also believed that *dark matter,* a force of energy and mass undetected so far, exists and makes up a large part of the universe. It is possible that developments in these branches of physics will lead to major shifts in how we view the universe.

To trace all of the scientific movements and changes in each period is beyond the scope of this chapter. Science certainly moved through different stages after the scientific revolution. For example, modern globalization and the rapid advances in communication in the twenty-first century hastened communication and collaboration among different peoples. On the other hand, the decline in science literacy among the U.S. populace presents a new threat to science advancement, as discussed in other chapters.

Scientific development progressed throughout the postscientific revolution period. It is clear from the general chronology in this chapter that science occurred at different rates during different periods. In congruence with societal changes, science occurred as punctuated and accelerated at times, rapidly advancing in some periods and alongside lulls in other periods. One theme emerges—science cannot be stopped. This concept will be explored in another chapter.

## ■ KEY TERMS

alchemy

Aquinas, Thomas

Aristotle

astrology

atomic fission

Boyle, Robert

Boyle's law

Bruno, Giordano

calculus

comparative anatomy

Copernican revolution

Copernicus, Nicolaus

Crick, Francis

dark matter

*De Revolutionibus*

DNA (deoxyribonucleic acid)

Einstein, Albert

Enlightenment

ethics

Franklin, Rosalind

Galilei, Galileo

geocentric model

heliocentric model
Hermetic doctrine
Kepler, Johannes
metaphysics
microbiology
nanotechnology
natural history
natural philosophy
neoplatonism
Newton, Isaac
Newtonian physics
Paracelsus
Planck, Max
Plato

*Principa*
Ptolemy
radioactivity
quark
scientific revolution
Socrates
Socratic method
string theory
taxonomy
theory of relativity
Trismegistus, Hermes
van Leeuwenhoek, Anton
Vesalius, Andreas
Watson, James

## ■ PROBLEMS

1. Revisit Pascal's quote at the start of the chapter. How was it a symbol of the scientific revolution?

2. Why do you suppose the scientific revolution permanently changed the way science was thought about? Why was it not just a "fad"? When could the revolution reverse and under what circumstances?

3. Define the Aristotelian-Ptolemaic explanation of the universe. How did it differ from the thinking of Socrates? How did it differ from the findings of Copernicus?

4. Compare and contrast the following terms. State one way they are the same and one way they are different.
   a. Copernicus and Galileo
   b. Galileo and Bruno
   c. Newton and Kepler
   d. van Leeuwenhoek and Boyle
   e. Nineteenth- and Twentieth-Century Science
   f. Taxonomy and Nanotechnology

5. Explain what sociological changes led to the overturning of Aristotelian thinking.

6. Do you think that Plato would like his student Aristotle? Why or why not?

7. Which advancement in the nineteenth or twentieth century do you think was most important? Why?

■ REFERENCES

1. Nicolle, J. 1961. *Louis Pasteur: The story of his major discoveries*. New York: Basic Books, Inc.

2. Sherman, D. and Salisbury, J. 2011. *The West in the world*. (p. 452). New York: McGraw Hill.

3. Zeldin, T. 1994. *An intimate history of humanity*. New York: Harper Collins.

4. McKay, J., Hill, B., Buckler, J., Crowston, C.H., Wiesner-Hanks, M., and Perry, J. 2011. *A history of western society*, 10th ed. Boston: Bedford/St. Martin's.

5. Sherman, D. and Salisbury, J. 2011. *The West in the world*. (pp. 452–453). New York: McGraw-Hill.

6. Mayr, E. 1982. *The growth of biological thought: Diversity, evolution, and inheritance*. (pp. 84–87). Cambridge, MA: Harvard University Press.

7. Ibid, p. 25.

8. Sherman, D. and Salisbury, J. 2011. *The West in the world*. (pp. 451–453). New York: McGraw-Hill.

9. Ibid, pp. 454–456.

10. Ibid, p. 454.

11. Mayr, E. 1982. *The growth of biological thought: Diversity, evolution, and inheritance*. (p. 22). Cambridge, MA: Harvard University Press.

12. Sherman, D. and Salisbury, J. 2011. *The West in the world*. (pp. 457–462). New York: McGraw-Hill.

13. Ibid, pp. 458–466.

# CHAPTER 6
# Science and Society

## Factors Driving the Interaction Between Science and Society

The effects of science and its technological applications on social development are profound. As demonstrated in history, major changes in society's thinking—norms, styles, and control—depend on the direction of scientific movements. For example, developments of TV, radio, and the internet changed society in recent history. This chapter is dedicated to a look at the interaction between science and society. While it is obvious from the chapter on scientific history that science has great impacts on society, so too does society have influence over science.

What kinds of forces drive science? Clearly, economic factors drive research. Without money, it is far more difficult for science to advance. Money had to be made available for the development of the atomic bomb. Nuclear warfare had major psychosocial and geopolitical impacts on the world. The threat of assured destruction changed the ways of warfare, power structure among the nations of the world, and international angst about the future of humanity. It took money to do this.

But when did the atomic bomb reach full development? It was during World War II. There was the willpower of the people—the redistribution of resources, due to politics, to speed nuclear fission research and development. Would these developments have happened if it were not for the war? Ultimately, science and society inherently influence each other as depicted in **Figure 6.1**.

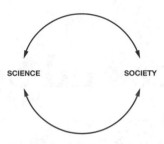

**Figure 6.1**   Interaction between Science and Society

## Positivist Perspective

Do scientists have an ethical responsibility for impacts their research studies have on society? Should the scientist be devoid of social forces in a pursuit of knowledge and truth? Is this even possible? In the discovery of evidence for evolution, should Charles Darwin have hid his thoughts for the sake of religious views? These are questions frequently asked in evaluating the ethics of science and its scientists.

History shows that science cannot be repressed, just as human thought and reason cannot be hindered. The proof is that science advances, sometimes slower or faster, throughout human time. The Dark Ages of the 1500s, a time of supposed loss in technological advance, also saw the growth of printing books and philosophy. Regardless of era, humans have a desire to explore and discover.

The *positivist perspective* holds that the role of the scientist is to discover "truth" objectively without concern for its impacts on society. In this view, no influence of society should contaminate the scientist's thoughts in theory building. It derives from the scientific philosophy of positivism discussed in another chapter. Without society, a scientist is able to pursue "true" and basic research free from the encumbrances of social rules.

The controversial Nazi scientist responsible for the development of German rocketry, Dr. *Wehrner von Braun* (1912–1977), once announced that, "Science does not have a moral dimension. It is like a knife. If you give it to a surgeon or a murderer, each will use it differently. Should the knife have not been developed?" Wehrner von Braun developed rocketry for Germany during World War II. That development led to the destruction of many lives, particularly in London.

The United States government not only waived Wehrner von Braun's accountability for war crimes but eagerly "denazified" him, and made von Braun a U.S. citizen. They appointed him head of the American rocketry program in the race to the moon. This scenario raises the provocative question: Should scientists be responsible for the results of their discoveries? Wehrner von Braun was not held responsible in the above situation. If the United States had not found utility in his scientific aptitude, would he then have been held accountable for the scientific developments that led to such

**Figure 6.2**  Sixty seconds after takeoff, the main engines on the space shuttle cut back to 65% to avoid stress on the wings and tail from the Earth's atmosphere. When the shuttle reaches thinner air at higher altitudes, the engines resume full power until the shuttle reaches orbiting speed.

grand chaos? And yet, von Braun laid the foundation for the development of modern rocketry, as shown in our rocketry to reach outer space in **Figure 6.2**.

## Social Constructivist Perspective

What is the role of scientists within a society? Are they realistically able to divorce themselves from the world? The *social constructivist perspective* contends that the role of scientists is to "construct" society and create science as a part of that society. In this perspective, scientists develop paradigms within societal norms. The role of the scientist is intricately linked with society and social norms because society creates the scientist—personality, education, philosophy, skills, and knowledge are all products of society. As such, the scientist creates and is created by culture. When a person develops a project or idea, social constructivism argues that it is the society that has led to that creation.

### Development of Evolution

Consider the development of evolutionary theory: Adam Smith's (1750s), Thomas Malthus' (1790s), and David Ricardo's (1820s) theories of monetary accumulation laid the foundation for the discovery of evolution. These economists argued that all societies tend toward a natural economic system of free markets and *capitalism*. This includes a freedom of people to compete for limited resources, with some being better suited than others, and thus a "struggle for the survival of the fittest" businesses.[1] *Charles Darwin* (1809–1882) author of the 1859 book, *On the Origin of Species*, grew up in an England that implemented these economists' views. The book delineated the evidence for changes in species due to competition. England in the 1800s was more purely capitalistic than any nation in the Western World today. Darwin's ideas developed in a society that functioned in such a way that it influenced his own hypothesis testing and theory building.

During Darwin's trip to the Galapagos Islands, he became aware that the different species of finches (a type of bird) had different beak shapes. These shapes were associated with different environments on each island. A drier island might have one shape for smaller nuts and a wetter island another shape. Thus, Charles Darwin deduced that at some point these birds must have competed for the limited resource of food. Further, this led to the development of different beak shapes to lessen the competition. He enumerated five steps to any species' development: (1) all species overpopulate, (2) individuals will then compete for the limited resources available, (3) that individuals of a population have variation (that is inherited), (4) some individuals have an advantage in their variation, and (5) a struggle for survival ensues, leading to "survival of the fittest" members of the population and thus a change in the characteristics of species over time. This is the theory of *natural selection* and thus the *evolution* of species, which mirror capitalism.

It is fair to say that society was "ready" for the theory of evolution. The capitalist world of thought had emerged and influenced Charles Darwin's thinking. Growing up in England gave him a mindset able to detect the patterns of data he observed and link it to organismal change over time. Society had influenced Darwin. If he had not grown up in a capitalist economy, his weltanschauung might not have enabled his inductive development of evolutionary biology.

Of course, true to the social constructivist perspective, controversy was inevitable. Religious views of the time held that humans were formed instantly by God and that humans were divinely special. Charles Darwin's views challenged the sanctity of man. A slow process of changing organisms, with humans just another rung on the ladder of evolution, displaced this purity. Darwin the scientist, in the face of opposition, recanted and revised his views, justifying the science in face of the social norms of the day. The stresses on Darwin show not only the influence of society on developing science, but also in suppressing it. That controversy continues for modern science, with the literal interpretations of the Bible and the rise of Intelligent Design theory as a competing alternative for the public. But, this is a good thing; debate and skepticism, as discussed in the chapter on scientific philosophy, are inherent in good science.

## Development of Homeostasis

While Darwin was influenced by capitalism, *Walter Bradford Cannon* (1871–1945) conversely developed scientific views influenced by communist economic theory.

*Communism* is the economic model that supports control over the means of production and consumption by a central government. It is a way to stabilize the economy by giving all citizens a portion of the wealth.

Cannon was essentially a communist in his outlook. He was also a Harvard physiologist and World War I medical doctor who pioneered the idea of *homeostasis*. Homeostasis is defined as the set of physiological (functional) mechanisms that maintain a "steady state" within living systems. Under homeostasis there is a centralized control region that sends out messages to other parts or systems within an organism to keep things in balance. In animal systems, a homeostatic mechanism regulates the body by first detecting a stimulus from the external environment. A receptor (or protein) picks up the stimulus and sends that information to a central control center (often the brain in animals). Based upon the control center's set point, which gives a particular value for the condition being monitored, a message is sent to effectors (any cell, often muscles or glands) to cause a response. **Figure 6.3** shows the flow of regulation in a homeostatic mechanism.

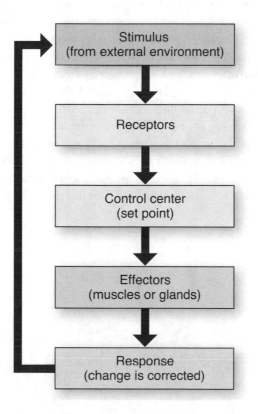

**Figure 6.3**  Homeostatic mechanism

For example, body temperature in humans hovers around 37°C (98.6°F). This is accomplished by the hypothalamus (control center) of the brain. When temperature changes, the hypothalamus sends out nerve impulses to blood vessels in the skin (effector). A temperature decrease causes *vasoconstriction,* where the vessels constrict (get narrower), restricting blood flow to the skin and conserving heat. The opposite happens when temperatures rise, with *vasodilation* (vessel gets wider) of blood vessels and more blood flowing to the skin to disperse heat.[2]

Another example of homeostasis is the regulation of blood sugar (*glucose*) levels, which normally measure about 90 mg/mL. The pancreas (control center) is the main organ of glucose regulation. When too much glucose is detected in the blood, the pancreas produces the *hormone* (chemical messenger) *insulin.* Insulin attaches to body cell receptors and this causes cells (effector) to take up glucose. *Receptors* are specially shaped proteins on cell surfaces that fit with certain chemicals in the body. See **Figure** 6.4.

Insulin in Figure 6.4 causes the cells to take up glucose from the blood. This lowers blood sugar levels. Conversely, when blood glucose levels decline, the pancreas makes the hormone *glucagon.* Glucagon works in opposition to insulin, raising blood sugar by causing the liver to change *glycogen* (the storage form of glucose) to glucose and releasing it into the blood stream. The two hormones maintain homeostasis through a process called *negative feedback.* During negative feedback, as one event occurs in living systems, another event checks it. In this way, balance is maintained.[3]

Upon comparison with communist economic theory, it appears that homeostasis works in much the same way as a centralized governmental control structure. Systems within the body control changes from the outside and inside environment to limit deviations. Communist systems use central banks, government laws, and regulations to control the speed at which an economy grows or shrinks. Note that the hormonal (endocrine) and nervous systems are involved in each of the examples given describing homeostasis.

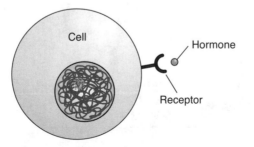

**Figure 6.4**   Insulin-Receptor Fit

Walter Bradford Cannon treated the shock and injury of World War I soldiers. He noticed that internal balances were disrupted when damage and surgery on the endocrine and nervous systems of the body were involved. Cannon furthered his research by removing parts of the nervous and endocrine systems to detect changes in these balances. He determined that various regions and systems of the body are under centralized control and termed this type of control *homeostasis*. Central controls (brain) used these systems to maintain order within the human body.[4]

Sir Bradford Cannon was most influenced in his development of ideas on human centralized control mechanisms from Italian economist Vilfredo Pareta (1848–1923), an anti-*laissez-faire* (free enterprise) proponent. Pareta was against free market capitalism and wanted a centralized form of government. The control of the economy by central government plays a major role in the United States today, in the form of the Federal Reserve and government assistance programs (e.g., Franklin Roosevelt's "New Deal" reforms). Cannon was influenced by these depression-era style politics. Thus it can be seen that society, namely pro-communist views of reform in the early twentieth century, influenced Cannon's search for "homeostatic" mechanisms.[5] The extent to which government should be involved in the economy is a major political issue today.

These are just a few examples of how society consciously and even subconsciously influences scientists. The direction science research flows is not only guided by the money and politics supporting science but also the social structure influencing the thoughts of scientists. These propel in line with psychosocial movements in the society. As described earlier, during an era of innovations in radioactivity, the politics of war sped the breakthrough of nuclear fission for application to the atomic bomb. The social constructivist perspective requires that the ethical role of the scientist be examined. Misconduct that scientists need to avoid due, in part, to the influence of money and society, is discussed in another chapter.

## Personal Responsibility and Conscience

What responsibility do scientists have for the uses of their research? Should they be held accountable for the applications of their results? We have discussed Wehrner von Braun. It is up to each student reading this text to determine the extent to which he or she believes scientists are to be held legally and morally accountable for their work.

*Robert Oppenheimer*, one of the contributors to the creation of the atomic bomb in 1945, upon viewing the first explosion of the weapon, commented that it reminded him of a line from the Bhagavad Gita, "Now I am become Death, the destroyer of worlds." As I wrote this textbook, the Japanese earthquake of 2011 was underway. Radioactivity from the nuclear plant in Fukushima plagues this nation again as it did once before following the dropping of the atomic bombs on Hiroshima and Nagasaki that ended World War II. The nuclear application of Einstein's equation relating matter and energy, $E = mc^2$ (where $c$ equals the speed of light) changed society in so many ways. Wars are fought differently, many nations use nuclear fuel for energy (30% of U.S. electricity is powered by nuclear power plants), and society both fears and values the possession of nuclear weaponry. What are Einstein's and Oppenheimer's responsibilities for changing the world?

After all, Rush Limbaugh, a contemporary conservative radio broadcaster, commented in 2005 that the Japanese economy benefited by the dropping of the bomb because it helped build up their economy after World War II. Is he correct in implying that we did the Japanese a favor? Applications and interpretations of events fostered by scientific advances vary broadly. Each individual is called to develop a personal perspective regarding the permissible extent to which scientists and we the public utilize known science. After reading this chapter one should be compelled to ask, "Where do I stand, morally?" on individual scientific issues. This text cannot dictate ethics but calls the reader to make very personal decisions based on the information.

## Ethical Dilemmas

An overview of all of the effects of science on society and vice versa is beyond the scope of this text. However, this text does seek to help people to explore the past, present, and future roles of science and to consider the possibilities. The main question is: What should we, as a society, do with the knowledge we get from science? Consider the effects of *prenatal testing* on the decision to terminate a pregnancy. Prenatal genetic testing, which screens developing embryos before birth, makes it possible now to determine whether or not a fetus will be born with certain genetic diseases. Examples include spina bifida, Huntington's disease, achondroplastic dwarfism, and cystic fibrosis. Recent developments in molecular testing are likely to make over 1,500 genetics screening tests soon available to the public. What if you were to find out your child would suffer from spinal muscular atrophy and die within 9 months of birth? Would you terminate the pregnancy? How about if the fetus is determined to have cystic fibrosis and is expected to suffer lung problems and die by age 20? What about Huntington's disease, in which the person is healthy until age 40 and then suffers progressive muscular weakness and dies by age 50?

Many younger readers draw the line against pregnancy termination here (because they think 40 years old is old enough!). What if you could determine your baby will get Alzheimer's disease at age 60 or suffer manic depression its whole life? What if your child were to have a form of dwarfism or congenital deafness? These questions are difficult and should be uncomfortable to consider. But we must all evaluate our moral choices. The answer may be "no abortion" to any of the questions posited or "yes" to all. Before technology made it possible to determine an unborn child's genetics, these questions would have been a nonissue. However, modern science has brought society to consider these difficult decisions. Of course, there is always the chance of error adding further difficulty to decision making.

Would you consider killing the baby after it is born and you discover it has a genetic disease? What if that discovery would have led you to abort the baby as a fetus? Most people would say "no way" because it is against the law. They are at the "authority" reasoning level in the *Typology of Argumentation* delineated in the chapter on scientific philosophy. Legally, this view is obviously supported; killing a newborn is murder. But, it is the legal system that sets the level of reasoning. Prior to prenatal testing, it was somewhat commonplace in ancient societies to abandon a newborn upon finding out about diseases. They had no ability to test the baby before it was born and this practice became a social norm. Human laws change with the discovery of new techniques such as genetic testing.

## Henrietta Lacks and Immortality

The best-selling book, *The Immortal Life of Henrietta Lacks*, by Rebecca Skloot, chronicles a unique case in which a woman's cervical cancer cells were harvested and grown in 1951, with the patient, Henrietta Lacks, dead from the disease that same year. Cancer is a disease in which there is abnormal growth of cells that leads to a takeover of other body areas. The cancer cells in Ms. Lacks were taken from her cervix by biopsy as part of her diagnosis and are still kept alive *in vitro* (in the lab) over 60 years later. In fact, a characteristic of cancer is that its cells are immortal. Given enough food they will live forever. They are often used for scientific purposes in experimentation and for observation.

As such, Henrietta Lacks' cells do not age or die like normal cells but remain, given nutrition, the same as they were in 1951. These cells are called the *HeLa cell* line because they come from the same set of cells from Henrietta Lacks when she was alive, hence the name. Recently, these cells were studied to determine the relationship between human papillomavirus and cervical cancer to develop a vaccine to prevent transmission of the cancer. The viral-cancer link was established for cervical cancer and is being marketed to immunize 9- to 14-year-olds for this type of cervical cancer. It was groundbreaking in that it showed that viruses may cause cancer. In fact, roughly 10% of all cancers are viral in origin, perhaps more.

Of course, cancer does die when the host organism dies, because it lacks nutrition from the body. Cancer cells are subject to the same kinds of needs as any other cell, but look and function differently from normal cells, with dedifferention (loss of specialization) and ability to spread (*metastasis*) obvious. **Figure 6.5** shows the development of a cancerous set of cells.

The major legal question is whether or not Henrietta Lacks was wronged by having her cells used without her permission in 1951. Do her heirs have rights to her cells for the scientific progress they have produced? There is currently litigation by her family

**1** An epithelial cell becomes partially transformed.

Partially transformed cell

Blood vessels

Lymph vessel

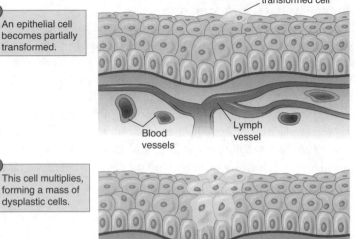

**2** This cell multiplies, forming a mass of dysplastic cells.

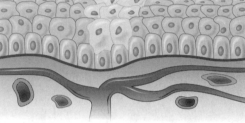

**3** These dysplastic cells grow rapidly, forming a localized cancerous tumor.

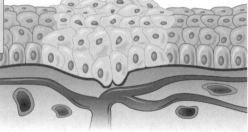

**4** The cancer cells secrete chemicals that allow them access to other tissues, the lymphatic system, and the blood stream.

Cancer cell secretions

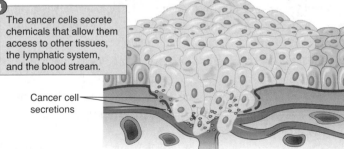

**Figure 6.5**    How Cancer Develops

members to gain the rights to those cells and the research that was produced. Human laws often attempt to control scientific discoveries and the impacts on society.

## Human vs. Natural Law

Natural law does not change with sociopolitical dynamics, however. Humans cannot control nature. We know that nature has tremendous power over human society. Earthquakes, global climate, and even the nature of human behavior are exemplary of the power of natural law. Consider the reflection by *Max Otto* (1876–1963), the science philosopher: "The universe is run by natural forces and laws, not by moral laws . . . Humans live by moral laws. If [they] contradict, it will be human society that suffers the consequences."[6] Human society continually struggles against the universe. Sometimes we are successful and use natural law for our benefit. Wind power, heart surgeries, farming techniques, and even nuclear fission are examples of scientific applications of human invention that better our lives. All too often, though, our efforts are thwarted. Tornados, hurricanes, and earthquakes are studied and detected but not avoided.

As humans, we tend to think of ourselves as organisms high in the evolutionary chain with much power and influence on Earth. To what extent do we truly have control over the planet? Let's look further at our control and impact on the weather. Since Earth's origin, there have been continual shifts in the climate.

Studies of ice cores from the polar caps show such fluctuations in modern geologic time. The *Little Ice Age,* an era of unusually cold weather taking place roughly between 1550 and 1850, after the Medieval Warm Period, had significant impacts on society in Europe and North America. This cold, a result of changes in solar output and a volcanic eruption, brought about starvation, desertion of Greenland's colony population, and even cannibalism in Europe. There were no significant anthropogenic (human-related) atmospheric emissions at the time of the Little Ice Age. Humans did not contribute to the climate change of the Little Ice Age.

However, today we see a strong correlation between global temperature and atmospheric carbon dioxide levels. We know our human activities are contributing generous amounts of carbon and sulfur emissions to the Earth's atmosphere via pollution. Are these anthropogenic emissions the cause of current global climate change? Do we as humans have the power and strength to willingly sway drifting global temperatures back toward a favorable direction, or are nature's machinations beyond our control? Global warming has become a geopolitical issue, wrought with human conflict between business and environmental interests. Sociopolitical influences from both sides have erupted into a battle that obscures the search for truth. Certainly, in either case, it is wise to limit frivolous atmospheric pollution (conserving energy and limiting waste where possible and economically feasible). The purpose of this text is not to take sides but to encourage us to question and investigate freely.

## Proximate vs. Ultimate Causation

In studying natural processes, the mechanism by which science occurs is termed its *proximate cause*. For example, the "what" and "how" of the atmosphere's chemical composition is its proximate understanding. Proximate questions are those that seek definable answers to natural happenings. What changes in atmospheric levels of gases are currently happening? What temperature changes are correlated with this? How does the Earth change its oceanic currents as temperature changes? These are proximate questions related to global climate change. Underlying questions about the larger meaning of science are termed *ultimate cause*. For example, the "why" and "reason for" the atmosphere's chemical composition is its ultimate understanding. Ultimate questions are those that seek indefinable answers to natural phenomena. "Why are humans on this Earth causing changes?", "Why do birds sing?", and "Why do humans love music?" are ultimate questions about natural phenomena.

While important, ultimate type questions, dealing with the "why" of science, can get in the way of objective research. Ultimate questions are answered by religion, philosophy, and sociology and often do not have a place in strictly interpreted scientific thought. At times ultimate questions may lead researchers to engage in a "groupthink" mentality and explore research areas tied strongly to their personal passions. As such, special care must be taken to remain objective. Scientists struggle to understand the universe and humankind's ability to find their place within it. However, in this tension, science is only able to attempt to objectively discover proximate causation.

The greatest struggle of all, of course, is our struggle within ourselves. It is human nature to want control over our lives—to have "free will." Or, are we merely nucleotides, a tiresome set of genes that govern us? Who controls our behavior, our genes, or the environment in which we are brought up? The scientific community has continued debate on the issue considering the influence of *nature* (genes) *or nurture* (our environment) on human behavior for centuries. Obviously the truth lies somewhere in between, depending upon the behavior considered.

A leading proponent of the nurture side of the argument was John B. Watson (1878–1958), who argued that the upbringing of a child was the only important variable in determining the outcome as an adult:

> Give me a dozen healthy infants, well-formed, and my own specified world to bring them up in and I'll guarantee to take any one at random and train him to become any type of specialist I might select—doctor, lawyer, artist, merchant-chief and, yes, even beggar-man and thief—regardless of his talents, penchants, tendencies, abilities, vocations, and race of his ancestors.[7]

Obviously there is a great deal of debate on the origins of human behavior. Consider the following case of scientific reasoning that debates the underlying causes of human behavior and the extent to which society or nature (our genes) controls us.

## Are Humans Inherently Good or Bad?

The assertion that "goodness" or "badness" even exists in animals and an animal social system is in and of itself controversial. It is therefore prudent to begin discussion of this issue by defining what is meant by good and bad. In the western world, *aggression* is considered both a good and bad characteristic. If it is applied appropriately (e.g., a person moves ahead financially, educationally) our capitalist system approves of and even glorifies aggressive behavior. The economically successful person is able to support his or her family and take care of more children, thus improving reproductive success. The individual is thus even considered *altruistic* in that she or he is working hard (decreasing the quality of her or his own life) to support and benefit others (usually family). Therefore altruism is conceived of as "goodness" in living systems. If, however, an aggressive act is perceived of as a selfish behavior, it is generally classified as a bad act. More specifically, a selfish behavior is defined as an act that benefits the individual committing the action but decreases the quality of another's life. If the same "good" individual described above (characterized as successful economically) has exploited his or her workers, that person's behavior would generally be considered bad.

Through studying the manifestations of aggressive behaviors, students may apply their ideas of goodness and badness in order to see the relative nature of the terms. The argumentation that underlies the extent of inherent virtue within living systems is clearly complex. However, when presenting the dialectics of opposing positions on the origins of human behavior, the agreement simplifies.

Konrad Lorenz (1903–1989) and the current Richard Dawkins are biological researchers in the study of human behavior. They offer viewpoints that expose their readers to an opposition of ideas about the inherent nature of goodness and badness both within humans and within their related social systems. The argumentation is fierce, drawing from examples of animal behavior to illustrate author points.

In the *selfish gene hypothesis,* Dawkins examines the biology of selfishness and altruism to present a case that humans and all life are extremely selfish by nature. His viewpoint is that all humans are so inherently selfish that society should ". . . try to teach generosity and altruism, because we are born selfish."[8] Lorenz argues that it is not humans but society that propels people to becomes aggressively selfish.[9] He points out that other animal social systems demonstrate restraint in their aggression. Our human inability to control violent acts between one another may be a commentary on our crowded capitalist system pushing humans into unnatural aggressive behaviors

Let us follow the argumentation to illustrate the formation of tentative conclusions from the question of whether humans are inherently good or bad. The reader should not expect to answer this question with certainty and resolve. Clearly it is debatable, like all scientific issues presented within this text. It demands a level of argumentation at levels 6 or 7 (as described in the *Typology of Argumentation* in the chapter on scientific philosophy), which allows the reader to accept that some questions require personal judgment to reach even uncertain conclusions.

When treating a controversial topic such as this, the author's perspectives should be cast in the light of their general alignment with larger classifications of thought. To illustrate, Dawkins' view of *individual selection* at the genetic level underlies his argumentation while Lorenz's view derives from a *group selection* approach. Individual selection is defined as the survival of the fittest individual, with the individual alone competing for limited resources. Group selection is defined as the survival of the fittest group, with the individual within the group and the group as a whole competing for limited resources.

Konrad Lorenz, like other group selectionists, views behaviors in terms of their *species-preserving functions*. Behaviors evolved to benefit the group by increasing the survival of related individuals. Certain species survived because they had the right amount of aggression against their enemies, so the behaviors are now a part of their social systems. Lorenz draws from animal systems to show that animals limit their aggression and have emotions such as grief and love. These ostensibly "good" behaviors demonstrate that living systems are good inherently and that humans, related to other animals, must also be good by extension.

Consider the species-preserving functions of aggression. It is better for a species if the strongest individual takes the territory or the female so that his genes can survive and dominate. In this way, for example, aggression among two males will enable the better genes to get passed onto the next generation. The weakest males, with their genes, are selected out. The weaker males are usually not killed in normal mate competitions between animals—only their genes are not passed onto the female and next generation.

In fact, aggression in animals leads to more of the individuals surviving. It leads to a more even spacing of territory and resources so that as many can survive in a given area as is possible. A more egalitarian system is set up that more or less maximizes the use of the resources in an area. Generally, since *intraspecific* aggression, or aggression between members of the same species, does not involve killing, the number of individuals remains high. Everyone gets a piece of the pie either by specializing on certain resources or using the territories at different times. Aggression also serves in defense of the young from predators. Birds always care for their young and keep constant watch while balancing a search for food. The animals thus demonstrate a respect for life and thus value individuals. Groups and individuals survive by behaviors adapted to limit aggression, according to Lorenz's perspective.

Group selectionists have also looked to social systems in animals that cooperate for the common good of the group. Insect social systems such as those found among the animal order *Hymenoptera* (ants, bees, termites, wasps) exhibit cooperative and what is termed *eusocial systems*. Sterile castes and a system in which an individual shows altruistic behaviors characterize eusociality. A bee will sting an enemy to save the group. This occurs despite the fact that the bee's organs will rip out of its body along with the stinger, and within a few hours the bee dies. Worker ants do not reproduce to have their own children but instead serve a queen master and help one another for the survival of the colony. Many ant species have defender ants with abdomens that explode with formic acid spewing forth onto their enemies to protect their colony. This kind of behavior is the epitome of dedication and altruism, as shown in **Figure 6.6**. Is it a misconception that such goodness exists? Should humans take an example of how to treat our fellow citizens from the *Hymenopterans*?

This is entirely plausible. In fact, it is a good example of other living creatures practicing the tenets of Judeo-Christian, Muslim, and Buddhist "goodness." Cooperation, altruism, charity, and acts of kindness are all characteristics of these eusocial groups and advocated by modern religions. The tenacity with which an ant drags a piece of food back to her colony is illustrative of the point. However, is there more? Why is the ant doing this? Why is she not taking the food and at least stealing a bite for herself? After all, doesn't she deserve a piece for all of her work? Why isn't she taking some for herself, especially if another sister is carrying a smaller piece and not doing as much work? Humans might think in this way but ants do not.

This depiction of animal social systems as benevolent and based on group benefits is staunchly criticized by individual selectionists such as Richard Dawkins. They argue that the behaviors are solely based on an individual's desire to pass genes onto the next generation. This comes down to the group vs. individual selectionist views of DNA. Group selectionists view DNA as a device used by organisms to deliver the next generation. Alternatively, individual selectionists view humans as shells that hold a very controlling genetic material that functions with a singular goal, reproducing itself.

Individual selectionists argue, for example, that there is a benefit to the individual when eusocial creatures cooperate. In fact, they argue the point of *haplodiploidy*: This term describes any social system whereby the queen gives *parthenogenic* (virgin) *birth* to all of the males in the colony. The males are *haploid* and have half the full amount of DNA. Thus, when a male mates with a new queen, the *diploid* female (full amount of DNA) offspring are 75% related to each other. This is because their father contributes all of his genes to his children so there was no random shuffling that occurs in non-haplodiploid species. The children are exactly identical as far as their father's genes are concerned. Their only differences are in terms of the mother's 50% probability of giving certain genes to her children (as in human reproduction). All *Hymenoptera* are haplodiploid.

Why is this important in determining whether eusocial insects are inherently good or bad? One might answer that eusocial individuals demonstrate good altruistic behaviors, so why is the cause of the behavior important? It is important because "goodness" or "badness" is determined by the motivation behind a behavior more than the behavior itself. The eusocial insects' motivation, or say, their genes' motivation, is

**Figure 6.6**    Ant Colony Cooperation

what matters to individual selectionists in determining whether an organism is acting altruistically or selfishly. Why do sister ants help each other and their queen so loyally? In terms of helping behavior, eusocial organisms are more likely to help one another than other organisms because of their high degree of relatedness. It pays to help one another out because by helping one another, eusocial insects are actually helping 75% of themselves (since eusocial insect females are 75% identical). Humans are at maximum (besides monozygotic twins) 50% related. Human parents and children are 50% related. Human sisters and brothers are 50% related. There is actually a mathematical equation predicting when altruism should occur based on mere genetic relatedness. In accordance with Maynard and Smith's equation to determine when helping behavior should occur, it is determined that eusocial creatures should help each other and are in fact not really altruistic; they are helping their own selfish genes.[10]

It is the genes that "want" to get to the next generation and survive. Therefore, they help whoever is as close to itself as possible. The more related the genes, the more the helping. The less related, the less the organism cares about the other creature. This is an extreme form of selfishness in that it is masked as altruism and helping, but in fact is based on a cold, calculated attempt to help one's own identical genes in the other creature. This is why a mother cares for her child, according to individual selectionists. This is why a brother cares for his sister. Thus it is seen that natural laws of genetics, and not human law, may have more impact on human society than we would care to admit.

## ■ KEY TERMS

| | |
|---|---|
| aggression | haplodiploidy |
| altruistic | HeLa cells |
| Cannon, Walter Bradford | hormone insulin |
| capitalism | *Hymenoptera* |
| communism | individual selection |
| Darwin, Charles | intraspecific |
| diploid | laissez-faire economics |
| eusocial system | Little Ice Age |
| evolution | metastasis |
| glucagon | natural selection |
| glucose | nature or nurture |
| glycogen | negative feedback |
| group selection | Oppenheimer, Robert |
| haploid | Otto, Max |

parthenogenic birth
positivist perspective
prenatal testing
proximate cause
receptor
selfish
selfish gene hypothesis

social constructivist
  perspective
species-preserving function
ultimate cause
vasoconstriction
vasodilation
von Braun, Wehrner

## ■ PROBLEMS

1. "Society was ready for Darwin's theory of evolution." Discuss this statement. Be sure to relate nineteenth century economic theory to the statement.
2. List the five steps of Darwinian evolution.
3. Compare and contrast the following terms. Give one way the terms are similar and one way the terms are different.
   a. Positivist vs. Social Constructivist perspectives
   b. Sir Bradford Cannon vs. Vilfredo Pareta
   c. Homeostasis vs. Disease
   d. Proximate vs. Ultimate causation
   e. Selfishness vs. Altruism
4. What two organ systems in the body work to maintain "homeostasis"? Give a biological example.
5. Based on the readings in this chapter, to what extent do you think humans are good to each other? Good for the planet?
6. Explain why human behavior is different from bees, ants, and wasps.
7. Who do you think best describes most human behavior: Lorenz or Dawkins? Why?

## ■ REFERENCES

1.   Allen, G. and Baker, J. 2000. *Biology: Scientific processes and social issues.* Hoboken, NJ: John Wiley & Sons.
2.   Marieb, E. 2004. *Human anatomy & physiology,* 6th ed. (pp. 10–11). San Francisco: Pearson/ Benjamin Cummings.
3.   Ibid, p. 10.

4. Allen, G. and Baker, J. 2000. *Biology: Scientific processes and social issues.* (p. 155). Hoboken, NJ: John Wiley & Sons.
5. Ibid, p. 156.
6. Otto, M. 1926. *Natural law and human hopes.* New York: Henry Holt and Company.
7. Bazzett, T. 2008. *An introduction to behavior genetics.* Sunderland, MA: Sinauer Associates, Inc.
8. Dawkins, R. 1976. *The selfish gene.* (p. 3). New York: Oxford University Press.
9. Lorenz, K. 1974. *On aggression.* Orlando, FL: Mariner Books.
10. Dawkins, R. 1976. *The selfish gene.* (pp. 316–323). New York: Oxford University Press.

# CHAPTER 7

# Scientific Rewards

**Scientists Are Their Own Society**
Peer Review
Rewards

**Becoming a Scientist**

## Scientists Are Their Own Society

Scientific research, as seen in examples in other chapters, does not exist in isolation, and there is a constant effort to work together. Scientists communicate and contribute in a subculture to produce advancements. The process is very much a social endeavor and scientists form their own societies. Such networks are often very specialized and close knit, held in solidarity by a united set of intellectual goals. But there is competition for the limited money available to fund their research. The scientific community therefore continually seeks a balance between communicating results with one another and competing for the limited monetary and publication rewards available in science. Oliver La Farge stated this of the scientific community in the following statement:

> Thus at the vital point of his life work [doing research] the scientist is cut off from communication with his fellowmen. Instead, he has the society of two, six, or twenty men and women who are working in his specialty, with whom he corresponds, whose letters he receives like a lover . . . in the keen pleasure of conclusions and findings compared, matched, checked against one another—the pure joy of being really understood.[1]

Within this subsociety, a set of norms exists. The associated rewards and punishments will be discussed in other chapters. John Ziman described these norms with the acronym CUDOS: communism, universalism, disinterestedness, originality, and skepticism.[2] First, *communism* describes a scientific society in which knowledge is the property of everyone.[3] Scientists share their information by publishing and presenting their data in public forums. In this way, ideas are exchanged, built upon, and improved or debunked. It is a communal process. Newton, the developer of mechanical physics, attributed his successes to this comradery in the comment,

"If I have been farther it is by standing on the shoulders of giants."[4] In industry and commercial research, this optimal is not always practiced. Scientific findings are not always shared. Results of scientific studies are often not published in order to keep data secret from competitors. However, most research does flow freely through the scientific community.[5]

Science should be shared liberally and have no borders. Ziman terms this second characteristic of science *universalism.* A scientist's religion, gender, nationality, or political persuasions make no difference to the interpretation, use, or quality of the results. Consider the use of solar panels in **Figure 7.1** providing electricity to small homes in less-developed countries. Use of modern technology for the betterment of society should not be limited by national boundaries.

Conversely, a scientist's past accomplishments should also have no bearing upon the acceptance or rejection of new results. To illustrate, the renowned chemist Linus Pauling, with an impeccable reputation in atomic structure development, could not convince his contemporaries that large doses of vitamin C had medical benefits. The scientific community objectively evaluated his evidence and rejected his claim.[6] Even today, the link between vitamin C and cold prevention is studied and conflicting evidence continues to emerge. It is fortunate that Pauling's reputation had no bearing or influence within the scientific community's positivist paradigm. The community has known cases of failure to respect universalism, as will be discussed in other chapters.

Courtesy of Roger Taylor/National Renewable Energy Laboratory

**Figure 7.1**   Solar Electricity. Solar Panels produce electricity in a thrid world village. In these homes in Brazil, solar panels on each home provide enough electricity for two indoor compact fluorescent lights.

Often, politicians fear and are frustrated by scientists. They fear what they do not understand and are frustrated that they should not control science advancement by reputation and/or manipulation. True science occurs for the advancement of knowledge and not for political or personal purposes. As such, scientists are *disinterested* in the results of a study. *Disinterest*, Ziman's third quality, means that the scientist should not have a vested interest in the outcome of the study. Politically, there is often pressure to produce a positive hypothesis or pursue a certain direction of research by authorities. It is highly unethical for a scientist to choose politics over reason. A politician in the media once asked the question, "What's with the science people? They don't cooperate, they don't act normally like others." To this I thought, "You fear what you do not understand." True scientists cannot be bought but base their decisions carefully on reason and objectivity. Thus there is an inherent mistrust of authorities and politicians in the scientific community. Political influences are usually nonobjective. Carl Sagan stated this simply, warning that one of the greatest commandments of science is: "Mistrust arguments from authorites."[7]

Fourth, scientific work seeks to advance new and unexplored areas of research or has *originality*. The first project of a Ph.D. (Doctor of Philosophy) degree in a subject area is called a *dissertation*. It is the first work establishing the student as a scholar because it is original research adding to some field of knowledge in the scientific world. Science dissertations present new facts, procedures, applications, or furthering of a theory or postulate. Scientific research that adds no new knowledge is looked down upon.

Finally, the fifth trait of the scientific community is *skepticism*. While most scientists are very critical of their own work, a community of scholars always reviews the work of their peers. This allows a certain level of watching and judging for any presented or published work to proceed. This process is institutionalized in the form of peer review. The feedback can be very useful for the researcher to view his or her own work through the eyes of others. In this way, it is more readily modified and critically evaluated in an iterative process. Few published works, including this one, are in the same form as their original composition. Ideas move back and forth between scholars to produce a better product. Critical evaluation can be tough to listen to for the scientist, but it is an essential part of the peer review process.

## Peer Review

An *invisible community* of colleagues in science exists to conduct, what is termed, a *peer review* process of emerging research. There are three formal ways to present new scientific information. First, journal articles are the primary conduit for new data announcements. The articles are *refereed*, meaning that they are judged by peers with expertise from the same subarea of research. If accepted, an article is deemed worthy of publication and is an honor for the author. Second, *presentations* are given at scientific meetings or conferences held by organizations or associations within a scientific discipline. This can be a valuable precursor step before submission of a journal article for publication. During a presentation, there can be useful feedback and discussion between the presenter and the audience. Third, *scholarly books* are written to integrate the refined ideas derived from presentation and publication.[8]

Additionally, informal review processes happen continuously in a loosely organized manner. Electronic communication over the internet has hastened and streamlined the sharing of scientific exchanges. Scientists gather online socially, formally, and regularly. This sociological change allows visual information sharing over the internet, immediate transmission of large amounts of knowledge, and brings the scientific community closer together for more effective collaboration. As early as 10 years ago this was not possible en masse. Still, informal interactions such as social events, job seeking, funding opportunities, and even gossip comprise an important part of scientific information sharing. In these ways, information is constantly evaluated in a peer review manner.

## Rewards

Formal peer review for journal and book publication, as well as research grant money, requires an inherent competitiveness. There is limited space for publication and there is always limited money so scientists necessarily compete with each other for these limited resources. This competition helps maintain quality and research standards. During the peer review process the peer reviewers are kept anonymous and likewise the author is unknown to the peer reviewer. This helps minimize some effects of academic jealousies and differences. The reward system is based on quality of the product and not politics.[9]

Nonetheless, the review process is still sometimes contaminated by competition. Often, peers know each other's works and at times jealousies can interfere with good science. For the most part, however, the reward system is just. Good scientists attain monetary awards in the form of higher salaries and grant awards, royalties from books or patents, and consulting work.[10]

Underlying these accomplishments is the need for *peer recognition*, which is respect from the science community. A scientist should not care about what others think of him or her on a personal level. It is the work that counts. However, a scientist should care about the recognition of his or her ideas by a coherent and logical set of science cohorts. In this way, the work can be properly reflected upon and thus scientific knowledge is furthered.

## Becoming a Scientist

Scientific interest begins in childhood. The excitement of a child's eyes when seeing her or his first butterfly or learning about how water moves is contagious. Children renewed my own passion about science, seeing it through their eyes.

Once in the education pipeline of school systems, fewer and fewer students remain interested in science as they progress. This problem will be discussed further in other chapters. Becoming a professional scientist begins with a focus on science in early schooling, likely majoring in science in college, and perhaps choosing to continue on in a graduate science program. In graduate school, a master's thesis is completed to earn a master's degree and a dissertation is completed to earn the Ph.D. While the originality of the dissertation work is itself important, equally valuable for the researcher are the skills derived from doing such a project. Organization, patience, library skills, writing, and higher-level analytical reasoning are needed for a successful thesis or dissertation defense. These abilities can then be transferred to future research for furthering science.

At times, a year or more of postdoctoral employment (called a *postdoc*) is sought to further the new Ph.D.'s research and/or teaching. Scientists may enter commercial (e.g., pharmaceutical), industrial (e.g., nuclear power plants), or government positions, which are generally well paid. Many enter academic fields (teaching) in college or university settings. An academic generally first serves an institution as an assistant professor for a probationary period (5 to 6 years in the United States). During this time, the professor concentrates on teaching, university, and/or community service and scholarship. Depending on the institution and interests of the academic person, a proportion of time is spent in each domain and divided accordingly. After the probationary period, tenure is earned. This is a guarantee of employment free from being fired for political reasons.[11] *Tenure* allows the professor to pursue intellectual thoughts even if they may not be socially acceptable. In this way, a positivist paradigm is able to be followed with a reduced fear of ramifications from authority. Scientists have a certain amount of freedom to pursue thoughts that stray from the mainstream thinking. As such, many scientific wild goose chases have been followed and resulted in discoveries of basic knowledge, such as radioactivity. These accidental discoveries might not have taken place without tenured faculty free to muse. There is much controversy about the merits of tenure, but freedom to pursue thinking outside of norms is its major benefit.

It is true that these descriptors make scientists a unique community, perhaps even cut off from the larger society. They share select and specific knowledge that only scientific equals can understand. Scientists shouldn't play by the rules of political or personal gain. They get jobs that often are eventually tenured and are secured through retirement. They work very hard and long in a cognitive capacity known only to their subgroup. To the world, they are not normal.

When General Groves assumed command of the Manhattan Project at Los Alamos, he was shocked at the work habits of the scientists working on the project. Instead of an ostensibly focused group of intellectuals, he found concerts, parties and dances, gossiping and neuroticism, along with odd and frivolous pursuits. Dr. Robert Oppenheimer, head scientist, had to explain to the General that this is how scientists worked under pressure.[12] Given the information in this chapter, it is no surprise. . . And yet, it is under these conditions that the atomic bomb was made. An enjoyable time is shown in **Figure** 7.2 below, with a group of scientists socializing during the very serious production of the atomic bomb.

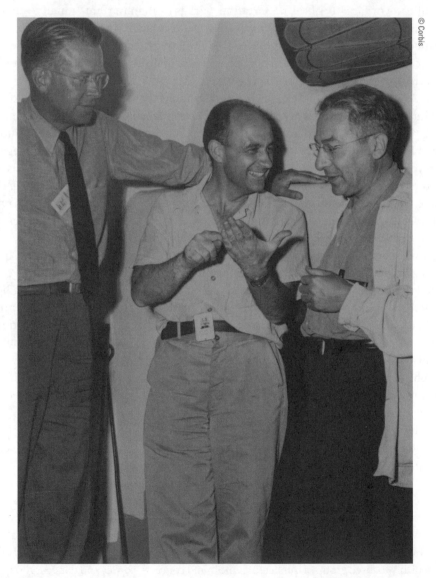

**Figure 7.2**    Los Alamos and the Manhattan Project

# ■ KEY TERMS

| | |
|---|---|
| communism | postdoc |
| disinterest | presentation |
| dissertation | refereed journal article |
| invisible community | scholarly books |
| originality | skepticism |
| peer recognition | tenure |
| peer review (formal and informal) | universalism |

# ■ PROBLEMS

1. Which of the five norms of the scientific community do you think is most important? Least important? Why?
2. Reflect on the statement by Carl Sagan, "Mistrust arguments from authorities." Give one example of where you found this to be true in your own life.
3. Trace the steps of becoming a scientist from the start of a college major.
4. Compare and contrast the following terms. Be sure to give one way they are the same and one way they are different.
   a. Formal vs. Informal peer review
   b. Publication vs. Presentation
   c. Dissertation vs. Postdoc
   d. Disinterestedness vs. Uninterested
   e. Originality vs. Universality
5. Why are scientists different from the larger community? Give three reasons. Do you think it affects their peer reviews?

# ■ REFERENCES

1. LaFarge, O. 1942. "Scientists are lonely men", *Harpers Magazine* (November): 652–59. Quote is on p. 657.
2. Committee on Science, Engineering, and Public Policy (U.S.) Panel on Scientific Responsibility and the Conduct of Research. 1992. *Responsible science: Ensuring the integrity of the research process,* Vol. I (pp. 5–6). Washington, DC: National Academy Press.
3. Ziman, J. 1984. An introduction to science studies: the philosophical and social aspects of science and technology. Cambridge: Cambridge University Press.

4. Merton, R. 1973. *The sociology of science: theoretic and empirical investigations* (pp. 274–275). Chicago: University of Chicago Press.
5. Lee, J. 2000. *The scientific endeavor: A primer on scientific principles and practice*. San Francisco: Addison, Wesley, Longman, Inc.
6. Hager, T. 1995. *Force of nature: The life of Linus Pauling* (pp. 573–627). New York: Simon and Shuster.
7. Sagan, C. 1995. *The demon-haunted world: Science as a candle in the dark* (p. 28). New York: Random House.
8. Crane, D. 1972. *Invisible colleges: Diffusion of knowledge in scientific communities*. Chicago: University of Chicago Press.
9. Lee, J. 2000. *The scientific endeavor: A primer on scientific principles and practice* (pp. 61–63). San Francisco: Addison, Wesley, Longman, Inc.
10. Ibid, p. 63.
11. Ibid, pp. 63–64.
12. Vandemark, B. and Harrod-Eagles, C. 2005. *Pandora's keepers: Nine men and the atomic bomb*. Boston: Back Bay Books.

# CHAPTER 8
# Scientific Integrity vs. Pseudosciences

**Honesty**

**Pseudoscience**
More Popular Pseudoscience
Less Popular Pseudoscience
Psychological Causes and Implications of Pseudosciences

## Honesty

The most important characteristic of a scientist is *honesty*. Science progresses only with the reporting of true data and research findings. The scientific community depends on shared information that is honestly reported and verified. There are a host of dishonest practices that some, and hopefully very few, scientists engage in. There are two general categories of scientific misconduct: fraud and plagiarism.[1]

*Fraud* is defined as a *falsification* or *manipulation* of data to produce a desired (but unsupported) result. As stated in other chapters, there is pressure on scientists to publish positive hypotheses and this encourages fraudulent data; it is perhaps the greatest threat to scientific progress. For example, research is extended based on previous findings. When those results are fraudulent, all of the future work on which it is based is also flawed. Many institutions ignore fraud in their research divisions due to fear of reputation damage.[2] It is the hope of this chapter to heighten awareness about systemic fraud and combat it, as shown in **Figure 8.1**.

While falsification, or outright making up of data, appears more dishonest than manipulation, they both lead the science community to serious problems. In the case of Stephen Bruening, he falsified results of studies on drug therapy for severely mentally disabled children. He falsely recommended stimulant treatment over tranquilizers. He reported an ineffective medical practice, which led to an endangering of the lives of thousands of children. In an attempt to further his career without doing the prerequisite work, Bruening exemplifies the damage to not only science but the greater community when the scientist places his/her gain over truth. A fellow scientist, Robert Sprague,

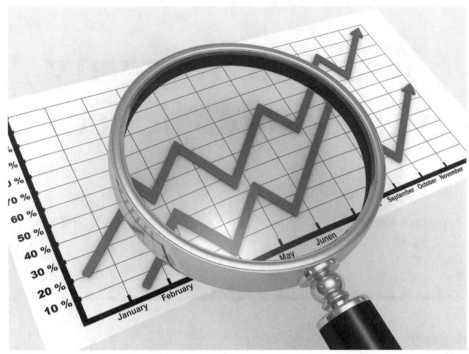

**Figure 8.1**   Fact Checker

discovered the Bruening fabrications and reported the misconduct to Bruening's funding agency.[3]

This kind of *whistleblowing* is very dangerous for the reporting scientist, like Robert Sprague. There are political ramifications. In fact, Sprague found himself under investigation and political attack for reporting Bruening. Thus, while it is the ethical responsibility for scientists to report *misconduct*, it should be done carefully and discreetly with senior staff and institutional support.[4]

The second type of misconduct, *plagiarism,* is defined as the use of someone else's words or ideas without giving that person credit. It is the equivalent of intellectual theft. Generally, when an idea is taken from another work, the source is cited so that the idea can be traced back by the reader to what was originally meant. Also, credit is given as a matter of respect and it establishes the history of the ideas. A direct quote must be cited with quotation marks around the quote and the source immediately given. Even a paraphrasing of an idea should be cited because it is coming from someone else.

Of course, there is a judgment call involved in deciding when to give proper credit. Some ideas like "bodies fall to Earth due to a gravitational pull" need not cite Isaac

Newton. The idea is common knowledge and as such it is assumed that we all know it. On the other hand, judging what "everyone" knows is sometimes unclear. As a golden rule: when in doubt, *cite* it. There are a host of other kinds of scientific misconduct, as reported by the National Academy of Sciences.[5] A few examples of note follow.

*Manipulation of public sentiment* occurs frequently when the results of a study are partially released or completely released without a full peer review. In the excitement of reporting results, some scientists and institutions announce findings to the media (which is all too eager to report) before they are ready. Consider the initial reports on *cold fusion*, which is a type of nuclear reaction occurring at low temperatures to yield large amounts of useable energy. In 1989, a breakthrough in cold fusion research occurred, in which chemists B. Stanley Pons and Martin Fleischmann held a press conference to announce their findings prematurely. They had not gone through the peer review process and had not yet been published in a reputable refereed journal. After about a year, other scientists could not replicate the research findings and it was deemed that Pons and Fleischmann did not actually achieve cold fusion.[6]

*Inappropriate use of mathematics or statistical design to interpret data* occurs frequently when scientists have more than one way to analyze data but fail to show ways that do not support their claims. As discussed in the chapter on mathematical analysis, this is deceptive and can lead to public health consequences. The tobacco industry was notorious in the reporting of only certain mathematical analyses in the 1950s and 1960s, downplaying the link of smoking with lung cancer. One out every nine people in the U.S. currently dies from a smoking-related cancer!

*Failure to give access* to proper and timely data reports, physical evidence, or procedural information is a grave violation of the principles of universality. Keeping a timely and meticulous lab notebook is crucial in any scientific investigation. It reports the process, changes in the procedures, and the thoughts of the scientist. Some agencies have requirements for the length of time data should be stored for public review after a study is published.[7] For example, the U.S. National Institutes of Health require that the written documentation of the findings contributing to one of their grants be kept for 3 years after publication or submission of the final report.[8]

*Conferring proper authorship or proper credit* for contributions to a published work is a judgment call by the principal investigators. In the graduate school world, many a student becomes the servant of their advisor or dissertation chair. Some professors are fair and just, giving proper credit to whomever gave useful contributions to their work. Others use ideas freely and take credit for themselves, again to more rapidly advance their own careers. In some cases, the credit to be given is unclear. Does a lab technician who performs only procedural tasks but does not contribute cognitively deserve authorship? Answers to such questions are based on past practice and ultimately the conscience of the author.

In all cases of scientific integrity issues, it is always the character of the scientist conducting the investigation that determines the honesty of the work. The science community is very small (in one's own area of research) and news travels quickly if there are questionable issues of integrity. I remember a Johnny Cash song with the line, "… bad news travels like wildfire, good news travels slow." If conscience fails, social pressure of the community often leads to prevention of scientific misconduct.

# Pseudoscience

If integrity is about personal accountability then an adherence to pseudoscience is based on society's proclivity to avoid that responsibility. In other words, people do not want to take blame for their problems and actions and instead look to pseudoscience to solve those problems. *Pseudoscience* is difficult to define but in general, it is a body of knowledge and methods presented as scientific but not grounded in or meeting the standards of scientific process.[9] Science is based on empirical evidence but pseudoscience is not. Science is verifiable and shows reliable results but pseudoscience does not.

Human beings have an inclination to blame happenings, both in their own lives as well as in the larger world, on something or someone other than themselves. This inclination is what draws the public to the variety of pseudosciences offered. Shakespeare describes this as the "… admirable evasion of whore-master man, to lay his goatish disposition on the charge of the star."[10] Shakespeare is alluding to the study of astrology, or making predictions based on star alignments. Many pseudoscientists make vast promises or explanations to their clients. In the time I have spent writing this paragraph, there were advertisements on TV for alternative medicines, astrology, and an old 1960s rerun of *I Dream of Jeanie* in which a séance was being held. Maybe I should shut the TV off.

Pseudosciences fall into two categories: those well integrated into society and those on the fringes of being socially accepted. I categorize them based on the look I get when their name is brought up. However, several U.S. polls by media outlets in 2004 indicate high support for most of the pseudoscience areas, almost on par with some religious views. Some results of the national polls show that over a third of Americans believe in ESP (50%), haunted houses (42%), ghosts (34%), and UFOs (34%); about a quarter accept things like astrology (29%), séances (28%), reincarnation (25%), and witches (24%).[11]

Generally *chiropractics* and acupuncture (alternative medicines with pseudoscience origins), fad diets, astrology, UFOs, ESP, and other paranormal phenomena appear to be somewhat accepted as valid by the general populace. Less acceptable but still seen with openness are creation science, dowsing, graphology (handwriting analysis), and cryptozoology (study of living creatures not identified by science; e.g., Bigfoot). The goal of this chapter is not to debunk or even debate each of these pseudosciences. Instead, it is to address the psychology behind a public acceptance of nonscience as science and to discuss this as a threat to scientific progress.

## More Popular Pseudoscience

Most pseudoscience is historically linked to the true sciences. Astronomy grew out of a scientific application of astrological data and modern medicine grew out of homeopathy and other philosophically based medicine of the times. This tie to science appeals to the public but the difference, again, is that scientific rigor is lacking. In looking into the various examples of alternative medicines rooted in pseudoscience, there is not significant evidence supporting many of the claims made. Alternatives to conventional Western medicine, while used for centuries in other cultures, have not been subjected to the same level of rigorous experimental designs as seen in the medical profession.

### Alternative Medicines

*Acupuncture* is the practice of inserting needles to stimulate nerves and treat health problems (**Figure 8.2**). *Small diameter nerve fibers* transmit pain but *large diameter nerve fibers* carry other sensory information. Scientists believe that acupuncture stimulates the large diameter fibers thus blocking small nerve fibers and pain messages. However, this is only a belief in how the process of acupuncture works and it has not been physiologically demonstrated. To illustrate, in some studies acupuncture has shown no significant improvement in patients given true acupuncture vs. those given

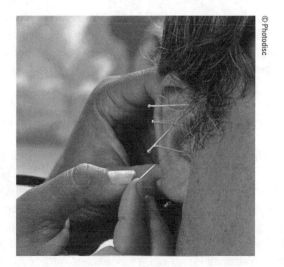

**Figure 8.2**    Old-New Treatments for Pain: Acupuncture

"fake" treatment. However, in other studies significant pain relief was shown. One, conducted by Joseph Helms, a physician with the American Academy of Acupuncture, performed acupuncture on 40 women experiencing menstrual pain, giving some real treatments and some placebo (needles placed in the wrong positions). The group receiving real treatments showed a 50% decrease in pain and the need for painkillers. However, the sample size of 40 subjects is small and a host of larger sample studies showed no difference in acupuncture vs. placebo.[12]

Alternative medicines, while favored by many patients for physical relief, advocate a philosophical rather than a scientific basis for healthcare. Acupuncturists, massage therapists, and chiropractors are licensed separately from medical professionals. Their philosophically based outlook for healthcare places them at odds with the science-minded American Medical Association (AMA), yet in recent years chiropractics and acupuncture have gained increased coverage and access under health insurance and HMO policies.[13]

Alternative medicine's healing claims are believable to the nonscientist because there is always the case, here and there, of someone who was helped by the alternative-medical treatment. There are two reasons for this. One is that many medical conditions improve on their own without intervention. This is especially the case for musculoskeletal problems (the mainstay of chiropractics) because inflammation and cartilage are changing and moveable thus resolving the pain. The second reason is that many conditions, like migraines (a common reason for acupuncture), are *psychogenic* in origin. If the patient "believes" he is being treated, then he starts to feel better. There are solid studies supporting chiropractic and acupuncture work but the body of research is much more limited than traditional medicine.[14] Indeed, these fields help many people but few medical practitioners work with or recognize these alternative medicines. Further research into alternative medicines will be required if integration into traditional medical science is to be achieved. Otherwise, alternative medicines need to remain classified as pseudoscience.

## Diets

*Diets* are perhaps the most common (and profitable) pseudoscience advertised to the public. This form of medical quackery is popular in a nation in which over half of its people are overweight. Numerous students and friends swear by different diets and their lack of good science needs to be addressed. In the example *Fit for Life*, authors Harvey and Marilyn Diamond contend that eating separate macromolecule meals (proteins, lipids, or nucleic acids) is recommended because their natural combinations poison the body. This diet was well received by the public. Further research, however, showed that such diets would lead, over time, to nutritional deficiencies and health problems.[15]

There are many other diets that claim weight loss. A more recent popular diet is the Atkins Diet, which emphasizes eating fat and protein and minimizing carbohydrate intake. The data do show initial signs of weight loss and a benefit to diabetics and pre-diabetics but actual weight gain in the long term. The initial benefits are probably because individuals are watching what they eat. Also in the patients studied, increased blood lipid profile values (higher bad cholesterol levels) and higher blood pressure were linked to a longer term Atkins Diet.

The golden rule of nutrition is that eating less than you burn off will result in weight loss. Alternatively, eating more than you burn off will cause weight gain. In fact, 3,500 kcal of extra food intake causes a person to gain 1 pound (.45 kg) of body weight. The Atkins Diet is also interesting biochemically: While a person eats a low carbohydrate diet, the body will transform fats and proteins into the needed carbohydrates essential to life functions. No one is on a low-carb diet because the body will simply make more carbs from fat and protein. The body takes care of itself through *anabolism* (building up macromolecules) and *catabolism* (breaking down macromolecules) regardless of the proscribed food intake dictated by the Atkins Diet. Individuals beginning any new diet are always advised to consult with a medical doctor first. Correlations with arteriosclerosis and heart disease make some of these diets dangerous to an unsuspecting public.[16]

## Astrology

Another socially acceptable form of pseudoscience is astrology. This is the ancient and popular belief that the position of the moon, sun, and other stars at the time of one's birth influence and downright determine one's personality traits.[17] This pseudoscience is so entrenched into society that I was offered my first college teaching job—I found out after the interview—because I was a Taurus. I was thankful and accepted the position but realized that getting work took more than just good grades and teaching.

Two studies have really debunked this way of thinking. John McGervey determined the "sun signs" (the astrological period in which one is born, e.g., Taurus or Libra) of thousands of politicians and scientists. He found no statistical tendency for members of either profession to be born under an expected sign more than any other sign.[18] Shawn Carlson conducted a controlled experiment on astrologers in which famous astrologers were asked to correctly match the personality traits of a subject with their astrological sign. Results showed no statistical accuracy in predicting a person's sign based on their personality.[19]

## Psychic Prediction

Frequently seen today, another commonly accepted form of pseudoscience, *psychic prediction*, claims to be able to foretell future events. Our accepted understanding of physics makes transfer of information from the future impossible.[20] Yet, people like *Nostradamus*, the great psychic predictor, are popularized often to have predicted this and that new event. Nostradamus's poems of the sixteenth century are unclear and vaguely written so that he allows almost any event to have been predicted by his writings. He was a clever charlatan, foretelling world history events using certain verbiage open to interpretation. One of the poems supposedly foretells of the fall of Germany at the end of World War II:

> Animals ferocious with hunger will swim the rivers,
> The greater part of the armed camp will be against the Hister;
> The great one will be dragged in an iron cage,
> When the German child watches the Rhine.[21]

While superficially alluding to the march into Germany by Russian forces in 1945 and "Hitler" being named, the verse needs to be more clearly looked at within the *historical setting* of the writing. At the time of the poem, the Turks were attacking but the French and Germans were suspicious of each other. Thus, the Germans were watching the Rhine River border with France but the Turks were a threat on the southern border, the Danube River. "Hister" is the Roman name for the lower Danube, where the Turks threatened, and does not at all refer to Adolf Hitler.[22] Unfortunately, the public neither gets exposed to a more reasonable and historic interpretation by the media, nor is inclined to accept it. Instead, Nostradamus is commonly presented as a gifted fortune teller.

## Less Popular Pseudoscience

A less socially acceptable form of pseudoscience is *dowsing*. Dowsing is the practice of finding underground water for well digging using some sort of forked instrument or stick. Why discuss dowsing? Freshwater is one of the scarcest resources in most of the world. Nations and cities are built upon the presence or absence of water for their development. The U.S. has an abundance of water, with 21% of the world's supply of freshwater contained within the Great Lakes alone. It is not difficult to understand how limited water is on the planet. Dowsing becomes a very attractive technique to many looking for this limited resource.

The dowser holds a stick, walking around, until the stick moves, seemingly uncontrollably, over the site where water is to be found. Many wells have been discovered with the use of dowsers and people swear by their efficacy. However, when dowsing is subjected to systematic scientific testing, it fails.[23]

A main reason for the success of dowsing in certain areas is that the underground water table is distributed relatively uniform under the ground's surface in many areas. Evon Vogt and Ray Hyman give an example of a 100% success rate for dowsers in a certain county in Alabama. Expectedly, the groundwater is the same depth throughout that county and the rate of success for non-dowsed wells is also 100%.[24]

In general, water wells drilled without the advice of a dowser are equally as successful as with their help. L. Keith Ward found that of 1823 doused and 1758 non-dowsed wells drilled in New South Wales, Australia, 14.7% of the dowsed wells were unsuccessful while only 7.4% of non-dowsed wells were. Thus, dowsing had twice the failure rate of random drilling.[25] Movement of the stick is either a hoax or due to subconscious control of muscles excited on the dowser when "thinking" a watery area has been discovered.

## Psychological Causes and Implications of Pseudosciences

Pseudosciences aforementioned in the chapter show common themes of direct appeal to the public without obtaining the evidentiary support of the scientific method. They do not use the rigorous peer review process, which establishes the truth and integrity of the research. It is a kind of scientific misconduct that is so base that it thrives on the psychological weaknesses of the human condition.

People seek a simple explanation for problems. In the face of lacking knowledge about a topic, the public looks to authority for answers, as discussed in the *Typology of Argumentation*. This lower form of reasoning is devoid of an ability to critically think about the topic. That is, the pseudoscience audience lacks the skill to evaluate evidence and draw reasoned conclusions. A certain amount of knowledge and reasoning in science, also termed *scientific literacy*, is lacking among the public, which allows pseudoscience to thrive.

The importance of critical thinking and ways to improve it will be discussed in another chapter. Without such abilities, pseudoscience draws on a *groupthink* mentality. In groupthink, the more people accept an idea put forth by pseudoscience, the deeper the beliefs are ingrained into the culture. There is an old adage to this kind of propaganda: If you repeat a lie often enough, people will begin to believe it. The greatest threat to the pseudosciences is objective truth and scientific reasoning to support that truth.

Unfortunately, there is a great deal of money to be made in advertising something masquerading as science. Both P. T. Barnum (circus promoter) and H. L. Mencken (writer) are said to have mentioned ". . . the depressing observation that no one ever lost money by underestimating the intelligence of the American public."[26] The remark has worldwide application. It is not, however, the result of a lack of public intelligence, but a lack of scientific literacy (knowledge of science and reasoning). It is not the fault of the people but of the mechanisms keeping the public from scientific skills, as will be discussed in the chapter *Roadblocks to Science*.

The ability to judge both evidence and alternative (or counter) evidence is an important critical thinking skill. One reason pseudosciences are so popular is that "people tend to believe what they are told, even when it comes from unreliable sources."[27] Many pseudoscience books and news reports fail to describe the counterevidence to their claims. Pseudoscience supporters even embellish stories to make claims such as UFO sightings and miracle weight loss more appealing.

Thus, developing skepticism is vital in cultivating scientific thought and combating groupthink.

Many pseudosciences make claims based on popular or political demand. Throughout the twentieth century, propaganda was used by many governments to mold public opinion in support of various ideologies and political objectives. Radio broadcasts, posters, films, print publications, exhibits, and educational and cultural exchanges were all part of a broader program designed to manipulate public opinion.[28] Albert Einstein, on concern for the lack of freedom of thought in the totalitarian world of the era, made a point of this in the following poem:

> By sweat and toil unparalleled
> At last a grain of the truth to see?
> Oh fool! To work yourself to death.
> Our party make truth by decree
> Does some brave spirit dare to doubt?
> A bashed-in skull's his quick reward.
> Thus teach we him, as ne'er before,
> To live with us in sweet accord.[29]

The ramifications of this groupthink mentality is a significant threat to modern scientific progress. Individuality and skepticism drive science and creativity to unlimited bounds, as seen by Einstein's discoveries. The push for individuality is a main thesis of this text. Many pseudosciences may one day show validity upon further scientific testing. Our minds must remain open. However, people taken in by the popular and the superficial will not be allowed to grow intellectually and will be flummoxed by the myriad of pseudosciences available.

## ■ KEY TERMS

acupuncture
anabolism
catabolism
chiropractics
cold fusion
diet
dowsing
fraud
groupthink
historical setting
honesty

large diameter nerve fibers
misconduct
Nostradamus
plagiarism
pseudoscience
psychic prediction
psychogenic
scientific literacy
small diameter nerve fibers
whistleblowing

# ■ PROBLEMS

1. Choose one of the pseudosciences. Do you believe what it claims to do? Form a plausible scientific investigation to test it.
2. Discuss the relationship between the following terms: falsification, manipulation, and whistle-blowing. Why is there danger in each of these?
3. How might the media change, in accordance with the National Academy of Sciences findings, to more accurately report scientific research?
4. Research a particular pseudoscience of interest to you. Draw a line down a page and list, on one side, *support* for the claims and on the other side evidence *against* the claims made by the chosen pseudoscience.
5. Reflect on the P. T. Barnum quote: "There's a sucker born every minute" in terms of belief in pseudoscience. How do you *best* think such an observation can be fought?
6. Do you think the origins of *groupthink* are genetically and/or evolutionarily derived? Explain your answer using Darwin's model of natural selection from the chapter on science and society.

# ■ REFERENCES

1. Lee, J. 2000. *The scientific endeavor: A primer on scientific principles and practices* (p. 66). San Francisco: Addison Wesley Longman, Inc.
2. Bell, R. 1992. *Impure science: Fraud, compromise & political influence in scientific research.* New York: John Wiley & Sons.
3. Lee, J. 2000. *The scientific endeavor: A primer on scientific principles and practices* (pp. 68–69). San Francisco: Addison Wesley Longman, Inc.
4. Ibid, p. 68.
5. Committee on Science, Engineering & Public Policy. 1992. *Responsible science: Ensuring the integrity of research process,* Vol. 1 (p. 28). Washington, DC: National Academy Press.
6. Huizenga, J. 1992. Cold fusion: the scientific fiasco of the century. Rochester: University of Rochester Press.
7. Lee, J. 2000. *The scientific endeavor: A primer on scientific principles and practices* (pp. 44–45). San Francisco: Addison Wesley Longman, Inc.
8. Marcina, F.L. 1995. *Scientific integrity: An introductory text with case studies* (p. 47). Washington, DC: ASM (American Society for Microbiology) Press.
9. Lee, J. 2000. *The scientific endeavor: A primer on scientific principles and practices.* (pp. 102–105). San Francisco: Addison Wesley Longman, Inc.
10. Ibid, p. 102.
11. Associate Press, IPSOS. 2007. AP/Ipsos Poll: *One-Third in AP Poll Believe in Ghosts and UFOs, Half Accept ESP,* http://www.ipsos-na.com/news-polls/pressrelease.aspx?id=3694. Accessed July 8, 2012.
12. Chiras, D. 2012. Human Biology, 6th ed. (pp. 214–215). Sudbury: Jones and Bartlett.

13. Velikovsky, I. 1950. *Worlds in collision*. New York: Pocket Books.

14. Fontanarosa, P.B., and Lundberg, G.B. 1998. Alternative medicine meets science, *Journal of the American Medical Association* 280 (16):1618–19.

15. Diamond, H. and Diamond, M. 1985. *Fit for life*. New York: Warner Books.

16. Barrett, S. 1993. Weight control: Facts, fads, and frauds. In S. Barrett and W. T. Jarvis (Eds.). *The health of robbers: A close look at quackery in America* (pp. 191–202). Buffalo: Prometheus Books.

17. Lee, J. 2000. *The scientific endeavor: A primer on scientific principles and practices* (pp. 103–105). San Francisco: Addison Wesley Longman, Inc.

18. McGervey, J.D. 1981. A statistical test of sun sign astrology. In K. Frazier (Ed.). *Paranormal borderlands of science* (pp. 235–240). Buffalo: Prometheus Books.

19. Carlson, S. 1985. A double blind test of astrology, *Nature* 318:419–425.

20. Rothman, M. A. 1988. *A physicist's guide to skepticism* (p. 150). Buffalo: Prometheus Books.

21. Robb, S. 1941. Nostradamus on Napoleon, Hitler, and the present crisis. New York: Charles Scribner's Sons, p. 171.

22. Lee, J. 2000. *The scientific endeavor: A primer on scientific principles and practices* (pp. 114–115). San Francisco: Addison Wesley Longman, Inc.

23. Ibid, pp. 104–105.

24. Ibid, pp. 103–104.

25. Ibid, p. 104.

26. Ibid, p. 130.

27. Ibid, p. 131.

28. Hamblin, T. 2006. *Selling America: The voice of America and United States international radio broadcasting to western Europe during the early Cold War, 1945–1954*. Ph.D. dissertation, Stony Brook, NY: Stony Brook University.

29. Calaprice, A. 2005. *The new quotable Einstein* (p. 238). Princeton, NJ: Princeton University Press.

# CHAPTER 9
# An Age of Optimism

**Major Controversies in Science Progression**

**Short Timers**

**Science as Solutions**

**Rapid Advances in the Early Twentieth Century**

**Deceleration**
Less Optimism

**The Next Big Innovative Revolution**

## Major Controversies in Science Progression

Throughout history, humankind was beset with conflict whenever changes occurred in society. Advances in science, in particular led to major tensions within the existing social structure. Attempts to cover up, reign in, or persecute people behind scientific developments were always defeated. Excitement of science and its possibilities could not be held back in any era, as seen in its history. This was well expressed by Buddha in the statement, "Three things cannot be long hidden: the sun, the moon, and the truth." Scientific truths, from gravity to nanoscale technology, always take hold of a society. While the controversies caused by scientific developments discussed in the chapter *The History of Science* were serious, truth behind the facts always won.

Galileo fought the Church, natural philosophers combated society, and people battled each other over scientific thoughts through the ages. A battle between the Church and newer thinking led to Galileo's persecution during the Scientific Revolution. This hindered scientific advancement for a stretch. However, in modern times, generally religious organizations are a supporter of, rather than a deterrent to, science. Religious organizations often support education and are in agreement with many tenets of modern science and education.

A focus on religion as an agent of scientific tension may be outdated. Instead there are a host of other current threats to scientific progress as delineated in the chapter

*Roadblocks to Science.* But scientific thinking always seeks controversy because it overturns the existing structure of thought. Battles against science continue today. Science is rooted in controversy and will always be the center of argument in a society. Through time, however, scientific truth based on reason always prevails.

## Short Timers

The miracle of science is how rapidly it has progressed within such a short time period. Consider the age of the universe, estimated to be about 15 billion years. In terms of geologic time, humans have inhabited the Earth only at the very end of this time scale: no more than 90,000 years. A single individual's lifespan is no more than a century. Yet a person can accomplish a great deal in that short time. Aristotle, Newton, Galileo, and Einstein produced a wealth of scientific change and knowledge in their lives.

*Carl Sagan* (1934–1996) compared human time on the Earth with the long lifespan of the universe. If the beginning of time to the present is compressed into one *calendar year*, consider the following: 15 billion years = 1 year; 1 billion years = 24 days; 1 second = 475 years. Humans as a species would only have emerged in the last hour and a half of the year. All of modern human history would represent only a second of time in the year's lifespan of the universe. Even less, a human life, far fewer than 475 years, is but a fleeting moment. And yet, in light of this, how all the more wondrous our accomplishments appear. Some representative dates are given in **Table 1**.[1]

| Table 1 | Key Dates in Sagan's Calendar Year |
|---|---|
| Pre-December Dates | |
| January 1 | Big Bang (Start of the Universe) |
| January 6 | First stars ignite |
| January 26 | First galaxies form |
| March 13 | Formation of the Milky Way galaxy |
| September 1 | Origin of our solar system and Earth |
| September 16 | Oldest surviving rocks on Earth |
| September 21 | First stirrings of life |
| November 8 | Significant oxygen atmosphere |
| November 30 | Invention of sex by multicellular algae |

*(Continued)*

| Table 1 | Continued |
|---------|-----------|
| Late Evening, December 31 | |
| 9:00 pm | First hominids |
| 10:30 | Stone tools |
| 11:00 | Domestication of fire |
| 11:48 | Symbolic language |
| 11:54 | Homo sapiens appears |
| 11:58 | Seafarers settle Australia |
| 11:59 | Extensive cave paintings |
| 11:59:37 | Invention of agriculture |
| 11:59:46 | Rise of Sumer |
| 11:59:48 | Bronze Age begins; writing invented |
| 11:59:51 | Code of Hammurabi in Babylon; Middle Kingdom in Egypt |
| 11:59:52 | Moses leads enslaved Hebrews out of Egypt |
| 11:59:53 | Trojan War; Olmec culture; Iron Age; Carthage founded |
| 11:59:54 | Founding of Rome; Ashokan India; Chin dynasty; Periclean Athens; birth of Buddha |
| 11:59:55 | Euclidean geometry; Archimedean physics; Ptolemaic astronomy; birth of Jesus |
| 11:59:57 | Zero and decimals invented in India; Muslim Conquest |
| 11:59:58 | Mayan civilization; Sung Dynasty China; Byzantine Empire; Mongol invasion; Crusades |
| 11:59:59 | Copernicus; voyages of discovery; emergence of experimental method in science |

## Science as Solutions

The past century witnessed an age of rapid scientific advancement. People have become secure in their confidence in science's ability to better their lives and the lives of the next generations. Solutions to problems that have plagued people, from cancer and heart disease to earthquakes and even meteorite threats, are assumed to be within grasp. The public views science as the answer to life's problems. This is demonstrated in America's drive for buying technology, plastic surgery, medical cost increases, and demand for better, greener energy to name a few examples. However, as far back as 1893, *Thomas Henry Huxley* (1825–1895) embraced the new age of scientific confidence, stating:

> I see no limit to the extent to which human intelligence and will, guided by sound principles of investigation, and organized in common effort, may modify the conditions of existence … and much may be done to change the nature of man himself.[2]

Scientists confirmed Huxley's optimism with sustained products and services that catered to human needs. Major discoveries and practical goods changed people's lives. Washing machines, automobiles, planes, trains, vacuum cleaners and electrical systems made life easier and leisure time and travel more accessible.

Research advanced in many fields. Discoveries in chemistry and physics laid the groundwork for the advances seen in our twentieth and twenty-first century science. Much of the advancement was seen in the growth of the modern university system. In the first decade of the nineteenth century, *German universities* first offered natural science as a field of study. Schools such as the University of Halle-Wittenberg in Germany led the way in 1836 for science as a developed major. They advanced scientific knowledge in two ways. First, they promoted original research in fields of knowledge outside of what was already known. This was the birth of the modern Ph.D. and innovation in specialized domains of study. Second, universities trained scientists, professors, teachers, physicians, and other professionals who would spread scientific thinking through society.[3]

*Dmitri Mendeleev* (1834–1907), a Russian chemist, compiled a new chart of known *elements* by *atomic weight*. In the 1870s, *James Maxwell* (1831–1879) analyzed the relationship between light, magnetism, and electricity to lay the foundations for the technological and electronics boom of the twentieth century. Russian scientist *Ivan Pavlov* (1849–1936) studied the conditioned reflex in dogs, whereby he established that many of our human responses are simple mechanical reflexes, starting the behavior psychology field. In medicine, new tools such as the thermometer and stethoscope helped diagnoses of ailments by measuring human vital signs. The microscope allowed examination of body tissue samples and x-rays (1896) imaged dense internal structures. The *electrocardiograph* (EKG) (1901) traced the electrical activity of the heart to show underlying activity and diagnose causes of cardiac disease. Louis Pasteur (1822–1895) gave the explanation for how people caught infectious diseases through his experimentation called *germ theory*, discussed in the introductory chapter. This led to improvements in medical care, including sterile techniques such as use of antiseptics, face masks, rubber gloves, and hand washing during operations and patient care. *Florence Nightingale* (1820–1910), a nurse during the Crimean War, transformed health care by stressing cleanliness, fresh air, and discipline. By the turn of the century, medicine gained the authority of the sciences and was "professionalized."[4]

## Rapid Advances in the Early Twentieth Century

A person born in 1900 witnessed stunning scientific progress and its applications. Between 1900 and 1950, people saw automobiles, telephones, television, airplanes, electricity, rockets, atomic bombs, radiation, jet propulsion, robots, computers, radio, and even Silly Putty for the first time. The first half of the twentieth century experienced the most rapid scientific proliferation of thought and creativity ever in history. Life changed from horses and no running water or electricity to the modern society that we now know. People could not travel by car or jet plane before this time period but the world used science to get interconnected. Rapid communications via TV and telephone, and transport using the gas-powered engine led to many societal changes. Life became much quicker.

Energy uses changed from merely wood and coal burning to uses of petroleum products as well as alternative type energy such as hydropower and nuclear fuel. **Figure 9.1** shows the relative proportions of energy options used in 150 years to present. Note that the proliferation of technological use of different sources of fuel emerges at around 1950. Pre-1950s science development laid the groundwork for these changes. While petroleum products remain a mainstay source of supply for the world's fuel, future technology will determine alternative fuel development and use.

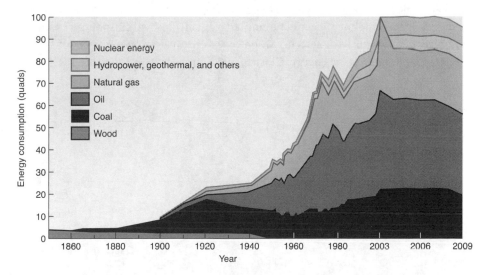

**Figure 9.1**  Changing Options. Energy consumption in the United States by fuel type from 1850 to the present. U.S. energy dependence shifted over the years from wood to oil, coal and natural gas.
*Source:* US Statistical Abstract.

Life also became more dangerous with technological advance, as weaponry became more rapid. Rocketry and nuclear threats of radiation permeated political policy after 1950. The Cold War and fear of annihilation dominated geopolitical thought. Radioactivity ushered in an age of uncertainty along with the age of optimism about science. It became obvious that science had consequences that could sow the seeds of human destruction. Nuclear radiation and technological advances in the military could produce a rapid destruction of civilization. This threat remains today.

The more people learned the less solid existing knowledge would become. Albert Einstein (1879–1955) developed his theories of *relativity*. Relativity states that time and space are relative, not fixed. Thus, motion of objects cannot be simply understood by space alone but must consider time as a factor. The time passage an object experiences varies with its speed. Newtonian physics was challenged by implying that both time and space were variables that could be altered in mechanical motion. For example, relativity stated that the rate of time itself passing could vary with the speed of an object. As an object approaches the speed of light, Einstein contended, its passage through time would slow down. Thus, if you left Earth at the speed of light and returned a minute later, Earth would advance many years into the future relative to your slowing down in time.

The *general theory of relativity* also inferred mathematically that matter and energy are interchangeable. Obviously, Einstein added a new dimension to Newtonian physics both by distorting the distinction between matter and energy and adding new aspects to the movement of objects. Relativity ushered in the nuclear era by uncovering that a small amount of matter could yield a large amount of energy. In his studies, Einstein surmised that atomic energy could be emitted from any known substance. As shown in **Figure 9.2**, *nuclear fission* occurs when a particle breaks down into smaller parts to release large amounts of energy. Close to 300 billion times as much atomic energy

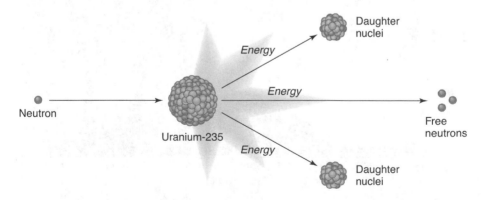

**Figure 9.2**    Nuclear Fission. In a fission reaction, a uranium-235 nucleus struck by a neutron plits into two smaller nuclei. Neutrons and enormous amounts of energy are also released.

could be released by a lump of coal through fission as compared with its normal burning in a stove. Challenges to scientific thinking led to an uncertainty underlying our modern day optimism.[5]

## Deceleration

The creativity and invention of a small handful of scientists, such as Einstein, led to tremendous change in the lives of many. People have benefited significantly from twentieth century advancements. However, most people today are merely operators of technology. They know how to use it but do not know how it really works. They are not inventors but passively learn what they need to know. While this is acceptable for practical workings in society, advancement occurs when understanding and building upon knowledge leads to new applications.

Science progressed rapidly in this way, in specific areas during the latter half of the twentieth century (medicine, computer technology), with certain people having extended understanding in these areas. The majority of the public, however, was left out of the advancement in knowledge. In fact, scientific literacy among the population has declined over the same period.[6]

The innovation seen in the earlier part of the twentieth century is not evidenced post-1950. There were certainly scientific advances made in more recent times: flight to the moon, medical breakthroughs, and computer improvements. However, they were all improvements of prior technologies and knowledge. Since 1950, advancement has come primarily from old technology that has merely been improved upon. For example, we went to the moon, but rocketry was developed during WWII in the V1 and V2 styles. Medical breakthroughs use radiation, echo waves, and imaging that dates back to developments in sonar and atomic warfare pre-1950. Computers were around before WWII. Automobiles are more efficient and safer than in the past but we still travel with pre-1950 technology—the automobile.

As a child growing up in the 1970s, I watched *The Jetsons*, a cartoon about a future world. It had rapid, casual space travel, beaming people around for instant transportation, cities on other planets, and robots to do our work. There were no cars anymore, only mini space shuttles. The cartoon expressed optimism for science and its possibilities better than any other show. It expressed the sentiment of the people in the mid-twentieth century. While it was only a cartoon, it comes to mind that we have none of the advancements imagined. **Figure 9.3** depicts a possibility for the imagination of what might have been or might yet be.

### Less Optimism

Automobiles are bigger and more gas guzzling than ever as the SUV (now one-third of all vehicles) has taken the place of regular cars. Transportation is accomplished

© Angela Harburn/ShutterStock, Inc.

**Figure 9.3**   A World of Hope

through the same means as in 1950: cars, trucks, roads, trains, and airplanes. The vehicles are more refined but there are certainly no mini-rocket ships or teleporting. Nuclear warheads still threaten our shores regardless of whether we ignore the situation. No missile defense shield was ever effectively developed. Computers are the latest obsession. They were developed pre-WWII. They are smaller and more efficient but again, not a new innovation post-1950. In fact, the U.S. recently gave up its NASA space shuttle program and any plans for returns to the Moon. This is concerning to the progress of a civilized society.

The case for a deceleration in innovation and scientific creativity since 1950 is made not to insult modern society but to underscore the importance of continued focus on science as a solution. There are many threats to science. One major roadblock to its progress is elevation of the nonintellectual. Economic progress in the next decade will depend on the next scientific innovation. Learning and creativity need to be supported in the culture to develop new Einsteins and Galileos. This is the road to scientific progress.

## The Next Big Innovative Revolution

Future scientific developments do not yet have a name, are not taught, and have no impact on our economy because they have not yet been invented. They are still mental constructs of the future. Perhaps they are in the minds of the babies being born at this very minute. It is impossible to know what science will be like in the future. What we do know is how science developed in the past and what it takes for a society to develop innovations. We also know from history that scientific innovation always drives economic progress and military success.

I would bet you a nickel, but no more, that the next big society altering innovations will emanate from the fields of energy and biotechnology. They drive a large part of our economy now. How impressive it would be to develop a small powerhouse-style box to give electricity to each house. However, I am not altogether certain that these areas will yield the innovation to change our future. They will develop but have been slowly advancing for some time. Some other areas in which we might see the next big innovative revolution could be: robotics (consumer, commercial, and military applications), alternative energy automobiles, commercial rockets/orbiters, nanotechnology, cloning, artificial intelligence, and new weapons for the military.

Information technology and computer applications have made significant changes to society. The internet has improved the ability of industry to better manage inventory levels, ship items quicker, market its products to a larger audience, locate information quickly, and communicate around the world rapidly and more cost-effectively. Additionally the internet allows for a wide range of consumer uses such as shopping, gaming, communicating/socializing, organizing/storing information, and locating information. This barely skims the surface. Information technology has been a large part of the focus of our nation's science progress in the past decades. It may well be the face of the future. It may be surpassed by something more powerful.

These extrapolations are taken from our present knowledge base. However, there could be some discovery that, again, is not yet even imaginable. Is there anything that can be done to encourage the next great revolution? Possibilities may include: more government funding of universities and private companies, reducing regulations to allow more biotechnology research, or tax breaks to companies that hit certain research and development (R&D) benchmarks. However, central to all of these ideas is the development of a scientifically literate populace. Human capital is the most important asset to society's future in science. We require a critical mass of enough people in science to make these changes. No governmental policy will be more important in furthering science than a focus on the improvement of scientific skill in the population.

## ■ KEY TERMS

calendar year
electrocardiograph
general theory of relativity
German university system
Huxley, Thomas Henry
Maxwell, James

Mendeleev, Dmitri
Nightingale, Florence
nuclear fission
Pavlov, Ivan
Sagan, Carl

## ■ PROBLEMS

1. What were the benefits of the birth of the modern university system to science progress in the nineteenth century?
2. Explain the rapid growth of science in the twentieth century. What factors led to this advancement?
3. In Carl Sagan's "calendar year" of the universe, which date/fact most surprised you? Why?
4. Which technological advance most changed society in the twentieth century? Why?
5. Ultimately, do you see science as a solution or a curse in our future?
6. What technological advancements would you like to see in your lifetime? Which do you think will have the greatest impact to benefit society?

## ■ REFERENCES

1. Sagan, C. 1975. A cosmic calendar. *Natural History* December: 70–73.
2. Sherman, D. and Salisbury, J. 2011. *The West in the world.* (p. 462). New York: McGraw-Hill Publishing.
3. Ibid, pp. 465–471.
4. Ibid, p. 662.
5. McKay, J., Hill, B., Buckler, J., Crowston, C. H., Wiesner-Hanks, M., and Perry, J. 2011. *A history of western society,* 10th ed. Boston: Bedford/St. Martin's.
6. Daempfle, P. 2006. The effects of instructional approaches on the improvement of reasoning in introductory college biology: A quantitative review of research, *Bioscene: The Journal of College Biology Teaching* 32(4): 22–32.

# CHAPTER 10

# Roadblocks to Science

## STEM Drives the Economy

Science is a fascinating and rewarding career with a central place in our modern economy and society. It is the driving force behind our national competitiveness. According to the U.S. Department of Labor, only 5% of U.S. workers are employed in Science, Technology, Engineering, and Mathematics (STEM) fields but they are responsible for more than 50% of our economic growth. STEM employees invent products that change our lives and create jobs.[1] Yet only a small percentage of people pursue careers in scientific areas. In fact, the rate of successful entry by first-year college students into science, and its related mathematics and engineering majors, has declined considerably over the past four decades. There are many reasons for this national trend.[2] Two purposes of this chapter are to address this problem by (1) delineating the challenges science faces and (2) recommending changes to improve science's advancement.

### Societal Challenges to STEM

First, in the larger culture, socially attractive *images of science* students and scientists are severely lacking. Sports heros, Hollywood stars, and even politicians are venerated while science is depicted by the media as "nerdy." However, it is true that very few people who try for a sports or Hollywood career actually make it (less than 0.01% or 1 in 10,000). In spite of this, the media create a culture of risk takers who would rather spend their time going for a long shot. In the process, people miss out on an education and possibly a career in a science area. As a result, science loses good talent to these schemes and the people lose out on a possibly stable and more certain career.

Second, this creates a society that *accepts* the emphasis on sports and entertainment over education. That has serious, negative effects on society and scientific recruitment and retention of good students. By forgoing education in the place of these long shots, students focus on the game or Hollywood as a means to a future. Time and focus is spent away from academics and that sets a tone for students. Science takes work and patience but mostly time. When the focus is taken off academics, there is attrition from not only the sciences but from college as a whole. Third, the lure of a media-enticing career in entertainment or sports is based on passing *fads*. People in these careers are hardly remembered. Consider that very few people still remember Barbara Stanwyck or Joseph Cotton, former movie stars of the 1930s. Yet everyone knows Albert Einstein and Thomas Edison. If people do not remember these scientists, they certainly know their creations: radioactivity, light bulbs, electricity, and their applications. Science touches eternity.

Focusing on trends and foregoing an education for passing fads is unfortunate. Efforts to retain and recruit good people in science are threatened by pop culture. Science needs to be "sold" as cool and profitable. People will then follow into it, accepting the cognitive work behind the discipline. The more recent sitcom *The Big Bang Theory* perpetuates the stereotypes of scientists as "strange" and outside of the normal culture. However, the show does bring science and science people into the mainstream by showing the human and fun side of the scientific community, as shown in **Figure 10.1**.

## International Comparisons

While I commend any sports team or sports scholarship that emphasizes academics, the problem is that the focus is still always on the sport. Academic pursuit is a means to an end, which is success at the sport and not a career linked to an area of study. This phenomenon is peculiar to the United States. It is rare to find someone going to college on a sports scholarship in a foreign country. This deemphasis on science and academics may well be contributing to a national decline in science and mathematics achievement. According to the many years of *Trends in International Mathematics and Science Study (TIMSS)* reports, the U.S. has steadily fallen behind other countries in science and mathematics scores, from first place in the 1950s to last among the Western nations today.[3]

**Figure 10.1**    The Big Bang Theory Cast Members

## Improving Science Education

The state of science education is in jeopardy because education and intellectualism in the U.S. has serious roadblocks aforementioned. Statistics show students lagging behind other Western nations in all academic areas. It is a complex issue, but to start, an increase in the *standards for science teachers* would help gain teacher respect and assure good knowledge in the faculty. Through accepting only the best academically prepared students into science teaching, a *reputation* that teaching is to be respected and that science is an entrance into a valued area would recruit and retain more students. Better academic standards for teachers should be supplemented with higher wages to both boost prestige of the academics and attract an elite force of teachers.

At the same time, a systemized redirection of students who are academically not suited for academics or a science education into a vocational or nonacademic path would improve *discipline in the science classroom*. Science is academically rigorous. While time spent away from focusing on learning hinders almost all secondary education areas, science is particularly hurt because the content is so demanding. The science classroom is a magnet for behavior problems due to the difficulty level and thus a poor fit for the academically uninterested or under prepared.

Behavioral problems take a great deal of time away from the classroom focus. By placing students on a more appropriate path, it would enable them to contribute more to society. These students may enjoy a faster route to employment and independence in a practical field. Finding alternative, work-related choices for such students is successful in other nations as well as in our vocational technical programs such as the Board of Cooperative Educational Services (BOCES) in New York State. Many students interfering with science learning are forced into the system because of practical and political decision makers. The alternate path ideas are challenged by these forces. However, in the face of sustained declines in U.S. science performance among its students, this approach would remove the variables of discipline and respect as contributors to the problems in science education.

## Critical Reasoning as the Main Threat to Science

### Defining Reasoning

The greatest challenge to the advancement of science in the U.S., whatever the cause, is a lack of the populace being able to critically reason. *Critical reasoning* is defined as an ability to work through problems and ideas and develop logical conclusions. The word is used interchangeably with terms such as: critical thinking, scientific reasoning, scientific thinking, and rational thought.

Unfortunately, after 12 years of education through to high school, the majority of college students lack advanced reasoning patterns that are necessary for successful achievement in undergraduate science courses. Students need to be able to reason through scientific studies and conduct investigations of their own to be successful in the sciences. If society does not emphasize these important characteristics in their young people, science literacy will be weakened and scientific progress is impeded.

The purpose of this next "recommendations" section of this chapter is (1) to discuss and define the various forms of critical thinking that are needed for success in the sciences and (2) to present the results of a review of studies of teaching practices found in science courses that claim to develop reasoning. Based on these results, the goal is to provide useful strategies to improve critical thinking skills.

An ability to critically consider phenomena is a key to success in the scientific world. It is a key to science literacy. Such an understanding is necessary for people to use science for everyday consideration. Being able to understand the doctor when he or she is explaining something, study scientific areas of personal interest, research media claims, learn about the natural world, and survive the changing technology as it emerges requires that schools and the individual focus on scientific reasoning skill development.

## Current Science Teaching and Reasoning

Although science faculty claim to advocate instructional methods that improve student critical thinking skills, limited research and change in most areas of science teaching has been documented.[4] This can be attributed to a variety of factors. According to Judith Glick, science instructors, especially at the college level, are often scientists who are untrained in instructional theory and practice. As a result, these instructors rely on the methods by which they were taught in order to develop a framework to guide their teaching. This is most often a traditional pedagogy, characterized by a rigorous adherence to content transmission and not the development of reasoning skills.[5] Science classrooms tend to have large lecture classes that reinforce passive roles for learners and so a special challenge exists to promote reasoning.[6] Also, the college science laboratories tend to be fact-laden and noninquiry based, with activities that act in opposition to the development of reasoning skills.[7]

There is a fear among college science educators that content knowledge acquisition would suffer if time were to be dedicated specifically to reasoning skill development during the lecture or laboratory.[8] It is also presumed by college instructors that post-secondary students, as adults, should be able to use critical thinking strategies independently after reading course materials and listening to lecture presentations. When students are unable to do this, the blame is simply placed on deficiencies in secondary level preparation.[9]

## Reasoning and Impacts on Science Learning

Because about 50% of first-year college students lack advanced reasoning patterns, there is a distinct way in which students learn science.[10] A lack of reasoning limits the ways in which learners process and then practice science and science thinking.

Educational researchers in reasoning processes **William Perry** (1913–1998), in 1970, and *Patricia King and Karen Kitchener,* in 1994, found that these entering college students are dualistic (right vs. wrong only) thinkers who are unable to evaluate an argument based on the strength of the evidence. A number of studies of empirical research will be discussed that outline the deleterious effects of a lack of reasoning ability on achievement in science courses and for science literacy, in general.[11]

The ability to judge scientific conclusions, as discussed in the chapter on philosophical science, is critical in understanding science. A lack of higher-order reasoning skills among students should be addressed by science teachers and science education research. Traditional instructional methods, such as lecture and textbook assignments alone, are not effective in developing reasoning in students.[12] The educational researcher *Jean Piaget* (1896–1980) argued that in the traditional lecture-based classroom the teacher is the source of all morality and truth, and "from the intellectual point of view, . . . [the student] accepts all affirmations issuing from the teacher as unquestionable . . ." so that the words are dispensed without the need for student reflection.[13] Thus, a static, unchanging, and factually based way of knowing is perpetuated. Piaget argued that this traditional method of teaching consolidates the egocentrisms found in childhood by simply replacing "a belief in self with a belief based on authority, instead of leading the way toward the reflection and the critical discussion that help to constitute reason and that can only be developed by cooperation and genuine intellectual exchange" to improve reasoning.[14] Thus, a major purpose for this section is to explore models of reasoning and the associated empirical research on the teaching methods that seek to improve critical thinking in science courses.

Why should a student entering into a science classroom study how teaching and learning in science takes place? Because understanding one's own thoughts and how one is taught science is as important as learning the science. There are challenges to thinking like a scientist and learning that information. The following section gives a scope of ideas on how science is taught and learned and how it is that we develop our own critical reasoning abilities throughout the college years.

## Theories of Reasoning Development in the Science Classroom

Most of the recent research on the teaching and classification of reasoning in science courses use Piaget's theory of intellectual development in interpreting their results. The Piagetian model of thought development identifies lower-level reasoning, called *concrete reasoning*, as being limited to merely describing and ordering of observable phenomena.[15] The concrete reasoner needs to reference familiar situations to accommodate and assimilate new information. For example, the concrete person must see or feel the item studied to understand it. Thus, only an inductive method of analysis (defined as reasoning from particular facts or situations to reach general conclusions) is employed to form conclusions. This type of reasoner lacks an awareness of thinking patterns and when faced with inconsistencies in evidence, is unable to generate or consider alternate hypotheses and so relies primarily on authority and intuition to draw conclusions. In the science laboratory, for example, this student is in need of step-by-step instructions during lengthy procedures.[16] As discussed in other chapters, this is the lowest level of argumentation in science and is exemplary of a lack of science literacy.

Higher-level reasoning, called *formal reasoning* in contrast, is characterized by the ability to generate and test alternative explanations when confronted with ambiguity. Much like Plato, this person plays with possibilities within the mind. Ideas such as math equations or problem solving can be thought out without seeing or feeling them, for example.

Reasoning is begun by imagining possibilities so that conclusions are drawn using, for example, the hypothetico-deductive method (defined as reasoning from a known general principle to the unknown) described in other chapters. These reasoners demonstrate the use of formal reasoning patterns. Formal reasoning includes the ability to control variables, and use probability, proportional, correlational, and combinational reasoning skills. This stage also involves *informal reasoning*, the systematic consideration of alternate hypotheses and evidence to draw conclusions. With such reasoning, individuals possess **metaknowledge**, or knowledge of their own thinking, and can thus evaluate inconsistencies in their own arguments. Piaget's formal reasoner is an independent thinker and can, for example, develop a workable plan of analysis in a science laboratory given the overall goals and resources of a lengthy procedure.[17]

This development of reasoning is related to the individual's ability to understand the nature and defense of one's own knowledge claims. As stated in the chapter on scientific philosophy, the area of philosophy that is concerned with the nature and justification of knowing is termed *epistemology*. A body of research exists based on how epistemological assumptions influence the development of reasoning. This includes, for emphasis in this text, the manner in which individuals come to know their own processes of thinking and reasoning.[18]

The ability to think critically in science is related to an individual's epistemological maturation. The progression of student reasoning abilities should continue through a series of stages from concrete to more pluralistic views, in which knowledge and values are perceived as relative. Perry defines the higher stages of reasoning strategies as being able to employ skills to interpret evidence to form conclusions. Thus, the student at this level, termed relativism, accepts the existence of possibly conflicting, multiple viewpoints and evaluates the evidence, internal consistency, and coherence of each perspective to formulate a conclusion.[19]

According to Perry's *relativist model* of intellectual development, higher levels of critical thinking involve a perception of knowledge and values as contextual and relativistic. Thus, in the science classroom, informal reasoning translates into skills in interpreting data and observations, evaluating equally valid arguments, and drawing conclusions from experiments. Accepting a number of possible answers is the road to this higher level. The intermediate level is termed *multiplicity*. Dualistic, lower-level reasoners are uncomfortable with the uncertainties involved in interpretation and evaluation of scientific evidence and so decision making in science becomes an incomprehensible process when the "right answer" is not provided. Thus, instruction should enhance student reasoning to relate scientific evidence with conclusions rather than simply focusing on memorization of those conclusions.[20]

Although Perry's scheme, influenced by Piaget, addresses general thought development, King and Kitchener point out that some aspects of scientific reasoning are not adequately described by either theorist. Thus, as an extension of Perry's work, King and Kitchener proposed a model that represents the most recent and extensive work on the development of informal reasoning in college students.[21] King and Kitchener conducted a 15-year interview-based study involving the analysis of reasoning in subject responses to *ill-structured questions*. Ill-structured questions are defined as having the possibility of more than one acceptable answer. Through this, King and Kitchener proposed a seven stage scheme for reasoning development called the *Reflective Judgment Model*, which focuses on the individual's understanding of the nature of knowledge and the process of reflecting on and justifying that knowledge.[22] Table 10.1 below compares the models of reasoning development described by Piaget, Perry, and King and Kitchener.

There are three levels within the Reflective Judgment Model seven stage model: pre-reflective (stages 1, 2, and 3), quasi-reflective (stages 4 and 5), and reflective (stages 6 and 7). In the pre-reflective stages, what is observed or what authority dictates determines truth. As with Perry's dualism, the individual is unable to reflect upon uncertainties in answering an ill-structured question.[23] During the *quasi-reflective* levels, there is a growing recognition that the individual cannot know with certainty and that each person is entitled to an opinion. It is during these stages that the belief that knowledge is relative emerges, yet the ability to actively construct arguments and evaluate scientific evidence is absent.[24] Only at the *reflective* stages, does the role of the knower move from a spectator and receiver of knowledge to an active constructor of

| Table 10.1 | A Comparison of Models of Reasoning Through Late Adolescence | | |
|---|---|---|---|
| Reasoning Level | Piaget | Perry | King and Kitchener |
| Low | Concrete | Dualism | Pre-reflective |
| Medium | Transitional | Multiplicity | Quasi-reflective |
| High | Formal | Relativism | Reflective |

meaning. Knowledge is recognized as uncertain and relative so that conclusions made from ill-structured questions include the critical evaluation of different positions. The highest level of reasoning in science occurs at this stage when the use of critical inquiry and hypothetical justifications allow for the evaluation and reevaluation of evidence and conclusions for ill-structured questions.[25]

The Reflective Judgment Model contends that reasoning abilities develop by assimilating and accommodating existing thought structures through interaction with the environment. The mechanics of this model of change are thus Piagetian and suggest that critical thinking should be emphasized in any science classroom or laboratory setting, where the opportunity for new thought processes are available.[26]

The higher-level reflective judgment characterizing stages 6 and 7 has been observed in only a tiny fraction of undergraduates interviewed by King and Kitchener and has appeared consistently only among advanced graduate students.[27] In addition, although it appears that education is positively correlated with reasoning stages, little development actually takes place during the college years. Less than half a stage of progression during the entire 4-year undergraduate experience is the norm reported by King and Kitchener. Thus, what kinds of teaching methods and learning environments would foster further development of reasoning in college students?

Further, what are the particular instructional variables within the methods that influence reasoning? Does course content achievement suffer when such methods are employed? What other learning variables are influenced by these instructional methods/environments? What are the relationships between those variables?

## Research Studies on Reasoning and Instruction

In this section, let's review that reasoning is separated into two constructs based on the three frameworks: *formal reasoning,* which includes control of variables, correlational, probabilistic, proportional, and combinatorial reasoning and *informal reasoning,* which includes the ability to explore nature, raise questions, generate multiple working hypotheses, and evaluate evidence to develop a logical argument.[28]

The purpose of this section is to present the results of research on different instructional approaches that seek to improve reasoning skills in science courses. Most of the studies that showed reasoning improvements were nontraditional, inquiry-based, collaborative approaches. The inclusion of writing, direct teaching of formal and informal reasoning methods, with sufficient length of time offering critical thinking instruction, and an inclusion of a collaborative approach were variables that elicited positive gains in students' reasoning development. Statistical analyses of the research studies were considered, as well as their experimental design, to screen the research and produce the model in **Figure 10.2**. The results of the studies, taken together, are a powerful statement on a need to change science education goals and methods to better

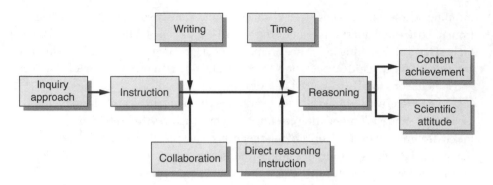

**Figure 10.2** Interaction of Instructional Variables on Reasoning Development

address the ways to improvement in critical thinking and hence science literacy for the population. How the instructional variables played a role in changing the reasoning patterns remains a black box.[29]

The following statements can be made with strong support from the empirical research of the studies reviewed, although they are not submitted without contestation. Several weaknesses were found in some supporting studies and there is not unanimous agreement on all points in the following recommendations for science education: (1) Inquiry-based, nontraditional *collaborative instruction* is more effective than traditional lecture instruction in developing higher order critical thinking skills in science courses. (2) The gains in reasoning through inquiry-based, nontraditional collaborative instruction are not achieved at the loss of *content acquisition*. (3) Inquiry-based, nontraditional collaborative instruction emphasizing *writing* to develop reasoning has higher success at developing a student's critical thinking than those methods not emphasizing writing. (4) The *direct instruction of formal and informal reasoning* leads to gains in those reasoning skill areas. (5) Gender and major do not appear to interact with instruction to influence reasoning. (6) Enough instructional *time* is needed to improve reasoning. (7) *Inquiry-based instruction* that improves reasoning also enhances positive scientific attitudes. (8) *Laboratory* is an important part of the science curriculum because it improves critical thinking skills when inquiry-based and collaborative. Figure 10.2 displays the purported relationships among the variables by the review.[30]

There are many questions that arise when considering the studies on reasoning more critically, namely the gaps in explaining how different instructional variables play a role in changing reasoning. However, it is clear that the research on reasoning, when taken as a whole, strongly supports use of nontraditional, writing- and inquiry-based, collaborative methodologies for the development of student reasoning.[31]

# ■ KEY TERMS

collaborative instruction

concrete reasoning

critical reasoning

epistomology

formal reasoning

ill-structured questions

informal reasoning

inquiry-based instruction

King, Patricia and Kitchener, Karen

meta-knowledge

multiplicity

Perry, William

Piaget, Jean

quasi-reflective

reflective

Reflective Judgment Model

relativist model

reputation

standards for science teachers

Trends in International
    Mathematics and Science
    Study (TIMSS)

# ■ PROBLEMS

1. Think about a science course you experienced in high school. Name three activities that could have improved the way students reasoned critically.

2. Compare the Reflective Thinking Model of Critical Thinking with Perry's views. Which do you think more accurately depicts reasoning skill development?

3. Why do you think that 50% of college students are at low-level reasoning patterns in science areas? What can be done to improve this?

4. Do you agree or disagree with Piaget's view that humans are locked into a stage of reasoning that cannot be broken through instruction? Why?

5. Why do you think writing is an important variable in reasoning skill development in science classrooms? Think of assignments you have had that really helped develop your thinking. In what subject was the assignment?

6. Do you think the media plays a role in turning students off to science? Why?

7. Do you think that sports play a role in benefiting or harming academics? Explain your answer.

8. Develop a science lesson on photosynthesis that incorporates the variables described in the model of reasoning improvement to address the suggestions found in this chapter.

# ■ REFERENCES

1.  Lehrer, J. 2011 (May 14). Editorial Why America needs immigrants. *The Wall Street Journal*, http://online.wsj.com/article/SB10001424052748703730804576313490871429216.html. retrieved July 10, 2012.

2.  Daempfle, P. 2006. The effects of instructional approaches on the improvement of reasoning in introductory college biology: A quantitative review of research. *Bioscene: The Journal of College Biology Teaching* 32(4):22–32.

3.  Ibid.

4.  Glick, J. G. 1994. *Effective methods for teaching nonmajors introductory college biology: A critical literature review* (seo54789). ERIC Document Reproduction Service No. ED373998. Available at www.eric.ed.gov.

5.  Daempfle, P. 2006. The effects of instructional approaches on the improvement of reasoning in introductory college biology: A quantitative review of research. *Bioscene: The Journal of College Biology Teaching* 32(4):22–32.

6.  Hall, D. and McCurdy, D. 1990. A comparison of a biolological sciences curriculum study laboratory and a traditional laboratory on student achievement at two private liberal arts colleges. *Journal of Research on Science Teaching* 27:625–636.

7.  Ibid.

8.  Sundberg, M., Dini, M., and Li, E. 1994. Decreasing course content improves student comprehension of science and attitudes towards science in freshman biology. *Journal of Research in Science Teaching* 31:679–693.

9.  Glick, J. G. 1994. *Effective methods for teaching nonmajors introductory college biology: A critical literature review* (seo54789). ERIC Document Reproduction Service No. ED373998. Available at www.eric.ed.gov.

10. Hofer, B. and Pintrich, P. 1997. The development of epistemological theories: Beliefs about knowledge and knowing and their relation to learning. *Review of Educational Research*, 67(1):88–140.

11. Lawson, A. 1992. The development of reasoning among college biology students—a review of research. *Journal of College Science Teaching* 21:338–344.

12. Hall, D. and McCurdy, D. 1990. A comparison of a biolological sciences curriculum study laboratory and a traditional laboratory on student achievement at two private liberal arts colleges. *Journal of Research on Science Teaching*, 27:625–636.

13. Piaget, J. 1970. *Science of education and the psychology of the child* (p. 179). New York: Orion Press.

14. Daempfle, P. 2006. The effects of instructional approaches on the improvement of reasoning in introductory college biology: A quantitative review of research. *Bioscene: The Journal of College Biology Teaching* 32(4):22–32.

15. Piaget, J. 1970. *Science of education and the psychology of the child* (p. 179). New York: Orion Press.

16. Allen, R. 1981. Intellectual development and the understanding of science: Applications of William Perry's theory to science teaching. *Journal of College Science Teaching*, 12:94–97.

17. Daempfle, P. 2006. The effects of instructional approaches on the improvement of reasoning in introductory college biology: A quantitative review of research. *Bioscene: The Journal of College Biology Teaching* 32(4):22–32.

18. Hofer, B. and Pintrich, P. 1997. The development of epistemological theories: Beliefs about knowledge and knowing and their relation to learning. *Review of Educational Research*, 67(1):88–140.
19. Daempfle, P. 2006. The effects of instructional approaches on the improvement of reasoning in introductory college biology: A quantitative review of research. *Bioscene: The Journal of College Biology Teaching* 32(4):22–32.
20. Ibid.
21. King, P. and Kitchener, K. 1994. *Developing reflective judgment* (pp. 2–15). San Francisco: Jossey-Bass.
22. Ibid, pp. 20–39.
23. Ibid, pp. 44–52.
24. Ibid, pp. 99–105.
25. Ibid, pp. 189–199.
26. Ibid, pp. 222–235.
27. Hofer, B. and Pintrich, P. 1997. The development of epistemological theories: Beliefs about knowledge and knowing and their relation to learning. *Review of Educational Research*, 67(1):88–140.
28. National Science Foundation. 1989. *Report on the National Science Foundation Disciplinary Workshops on Undergraduate Education*, 13–18. Arlington, VA: The National Science Foundation.
29. Daempfle, P. 2006. The effects of instructional approaches on the improvement of reasoning in introductory college biology: A quantitative review of research. *Bioscene: The Journal of College Biology Teaching* 32(4):22–32.
30. Ibid.
31. Ibid.

# CHAPTER 11

# Teaching Critical Thinking

**Improving Reasoning**

**What Is Critical Thinking?**

**Critical Thinking Strategies**
Affective Strategies
Cognitive Strategies
How to Think Critically in Our Society: Facebook Study
Media's Role in Critical Evaluation

## Improving Reasoning

The attempt to change teaching methods in science classrooms to improve critical thinking skills is not a recent phenomenon. An early study by J. Darrell Barnard at the Colorado State College of Education, in 1942 emphasized the need for students to learn more than just factual content.[1] The purpose of this chapter is to give critical thinking strategies; it is to engage the reader to use those strategies in activities both within the classroom and in everyday life. The goal is to aid the reader in improving the reasoning aspect of science literacy.

Reform efforts stimulated by *A Nation at Risk* in the 1980s to improve science reasoning have produced most of the studies on critical thinking in science discussed in the chapter on the roadblocks to science.[2] All of the studies were objective, quantitative, and experimental designs with an independent variable of teaching method and a dependent variable of improvement in reasoning. The most interesting theme emerging from the studies was the importance of writing and of collaboration during instruction to develop student reasoning. Many of the studies showed that integrating writing as an expression of reasoning during instruction had a positive impact on student critical thinking skills. Thus, this chapter asks the reader to engage in written critical thinking activities along with applying methods to improve one's scientific reasoning. This chapter is an excellent conversation piece for readers, encouraging discussion with others about salient issues brought forth in science. The topics and strategies presented should encourage discussions with friends and family.

Most of the studies reviewed in the chapter on the roadblocks to science implement Piaget's suggestion to ground the development of formal reasoning in concrete experiences and social interactions and through writing assignments. The end of chapter questions and activities draw from the suggestions on improving critical thinking by asking readers to write and express themselves in an evaluation of given information. Through linking relevant, everyday scientific questions, a mental disequilibrium may occur and reasoning may develop. This chapter asks readers to confront their scientific assumptions and then write about them. In this way, they are invited to bring science and perhaps misconceptions about science, into their everyday lives. Science and learning are social processes, so the inclusion of collaboration to discuss ideas by bringing these strategies to your classroom will also enhance your reasoning skills.

Findings in the previous chapter have major impacts on the future of critical thinking research in science. Future studies should explore the *relative* contributions of different variables within the model of instruction proposed in the chapter on the roadblocks to science that led to the successful development of reasoning. For example, the relative individual contribution of both the collaborative and inquiry methodologies should be determined. If a noncollaborative approach were used, how would the results on reasoning development change? Additionally, how would the introduction of a similarly intimidating, but nontraditional classroom environment change the results?

Further, could a student merely learn the models and steps of reasoning, but not actually be at a more sophisticated level? Evidence for this was seen in one particular study on reasoning by researchers Anton Lawson and Donald Snitgen in which there is no demonstration of the transfer of reasoning improvement to nonfamiliar topics and an actual decline in reasoning quality after teaching critical thinking strategies.[3]

This raises an important point: Is the teaching of critical thinking even possible in science, or any domain? Students could be intrinsically locked into a low Piagetian developmental stage of reasoning. The ability to change one's predisposed abilities before natural development allows for it may not be possible.

There are no studies explaining the mechanism for changing student reasoning skill—it remains a black box. Thus, although empirical results show increases in reasoning levels through the instructional variables suggested in another chapter, we do not know the causative agent. Is it an increase in neuronal pathways or a change in neurotransmitter levels in the brain that leads to critical thinking skill improvement? What is the nature of neuronal changes in students who improve their reasoning and students who do not? Regardless of the cause, it is the contention of this text, based on science education research, that instruction *can* positively affect critical thinking ability. The chapter goals are to improve reasoning through suggested methods and to apply those methods in examples and reflective questions.

## What Is Critical Thinking?

Simply defined, *critical thinking* is "the careful, deliberate determination of whether we should accept, reject, or suspend judgment about a claim—and with the degree of confidence with which we accept or reject it."[4] There are many terms associated with "critical thinking," as shown in **Figure 11.1**.

Other chapters defined the term loosely, using ideas including formal and informal processes, variable control, etc. The main determinant of critical thinking is the ability to judge a claim and change one's reasoning about it, if deemed necessary. It is human nature to hold biases and viewpoints based on one's upbringing and environment. The effects of "groupthink" and manipulation by authority are examples of integrity breaches in science described in other chapters. These are all antithetical to good critical thinking strategies and the advancement of scientific thought. However, there are ways to combat these forces. This chapter discusses a variety of strategies to critically evaluate arguments and thereby prevent such manipulation.

Many questions in science cannot be answered easily and require strategies to think them through. Critical thinking is a process of logical evaluation in which a definitive answer is determined after careful thought and research.[5] Often though, it is difficult for humans to engage in critical thinking; it seems to go against our nature. Wallace Stegner best describes this human proclivity in the following statement:

> One of the most difficult operations for imperfect mortals is the making of distinctions, of stopping opinion and belief part way, of accepting qualified ideas. But the individual who can modify or correct beliefs molded by personal interest or the influence of his rearing is rare. . . . It is easy to be wise in retrospect, uncommonly difficult in the event.[6]

The purpose of developing critical thinking in science is to expand the student mind beyond its basic confines and create a *scientific way of thinking*. Scientific critical thinking transcends the biases of one's surroundings and prevents integrity issues that cloud scientific progress.

## Critical Thinking Strategies

It is difficult to achieve all goals of critical thinking and reasoning under all circumstances. Richard Paul developed a list of strategies to best help us meet this ideal.[7] Paul divided his advice into two categories: an affective set of strategies and a cognitive set of strategies.[8] *Affective strategies* refer to those involving feelings, emotions, and attitudes. *Cognitive strategies* involve intellectual skills, such as interrelating ideas, memorizing, and developing hypotheses. In other words, cognitive strategies incorporate mental constructs. Each of Paul's strategies (shown italicized in the next sections) are discussed here in the context of science reasoning development.

**Figure 11.1** Reasoning Defined
© Kheng Guan Toh/ShutterStock, Inc.

## Affective Strategies

An ability to see the world from the outside in is described by Paul as both *thinking independently* and developing insight into egocentricity or sociocentricity. *Egocentricity* is defined as an inability to see another individual's perspective while *sociocentricity* is an inability to see the perspective of another culture or society. As a prelude to any course in science that requires critical thought, please read the short story entitled *Body Ritual Among the Nacirema* by Horace Miner. See an adaptation of the story in **Box 11.1**.

---

### BOX 11.1    TEXTBOX SUMMARY OF "BODY RITUAL AMONG THE NACIREMA," BY HORACE MINER

All human cultures develop their ideas about how to care for the body. The Nacirema society advanced rituals and rules of body maintenance to such a vast degree, that they spend a large percentage of their time, energy and resources doing so. These routines are passed down from generation to generation, with intense training of children on the need to care for one's hygiene. Professor Horace Miner (1912–1993) first brought the rituals of this culture to prominence, but the society of these people is still very poorly understood. Here is his account of what he observed while interacting with this tribe:

What is known is that they are a group in North America, living in the territory between the Canadian Tree, the Yaqui and Tarahumare of Mexico, and the Carib and Arawak of the Antilles. Not much is known about the exact origins of these peoples but tradition says that they came from the east . . .

Nacirema culture is based largely on a market system, which forces its tribal members to spend large amounts of time (over 2,000 hours per year) in pursuit of material items such as trinkets and treats for themselves. The rest of their time is spent in ritual activity focused on appearance and health of the human body in what observers call an obsessive manner.

The underlying belief system of this tribe is that the body is ugly and people are bad; it is only through obtaining fortune and beauty that a member of the society valued. Every household has a shrine devoted to this purpose. Man's hope in averting the deterioration of the body is to conduct ablution rituals in front of the shrine. Rituals associated with the shrine are secret and private, but I was able to establish enough rapport with the natives to examine the shrines and have their rituals described to me.

The center of the shrine has a chest with powerful potions expected to make people look beautiful and young. Medicine men provide curative potions for their people and they are looked up to and rewarded for their cures. The shrine has a charm box; and every day each member bows his head before it, mingling all sorts of holy water in the mouth and on the body. They conduct a rite of ablution in the charm box to ward off evil from their mouths. The natives hate their mouths and see them as prone to evil. There is a strong relationship between the mouth and one's social life—in fact, the mouth seems to play a major role in the number of friends and types of places one is allowed to go.

The private mouth-shrine ritual is accompanied by a visit to a holy-mouth-man once or twice a year to exorcise the evils of the mouth. The importance of the ritual allows the holy-mouth-men tremendous amounts of latitude in torturing the natives. In the end, the holy-mouth-men demand a gift of great expense for their acts. The population continues the torture each year at least, showing the masochistic tendencies of this culture.

The medicine men have a large temple, or latipso in each of the villages, in which ceremonies are held to treat the visitors. Some visitors never leave because of the harshness of the latipso ceremonies but if they do escape, charms are required of great value. The guardians of the latipso do not want to admit a client if the gift is not rich enough and pursue the person after he escapes if the gift is not sufficient, many times taking his home to compensate for a too small gift.

*(Continued)*

## BOX 11.1 CONTINUED

In order to cope with these troubles, there is a witchdoctor who has the power to exorcise the devils that lodge in the heads of people. The witchdoctor listens to the patient's troubles and fears, often together they blame the patient's mother, and some have a memory as far back to childhood, even complaining about the traumatic events of weaning as a baby.

The Nacirema are unhappy with their bodies and engage in all sorts of behaviors to change their bodies. There are rituals to make thin people fat and fat people thin; there are ceremonies to make breasts larger and breasts smaller. Excretory functions are hidden under all circumstances, except in the latipso, in which help is embarrassing but necessary. Pregnancy is hidden because there is a sense of shame associated with it.

On review, it is surprising that the natives just take all of this stress and suffering. It is brought on by themselves but nonetheless they have managed to exist for a long period of time. The exotic culture of rituals and crazy customs takes on a new meaning when viewed with the insight from Bronislaw Malinowski (1884–1942) when he wrote:

"Looking from far and above, from our high places of safety in the developed civilization, it is easy to see all the crudity and irreverence of magic. But without its power and guidance early man could not have mastered his practical difficulties as he has done, nor could man have advanced to the higher stages of civilization."

Reflection Questions:

1. Describe your initial thoughts about the Nacirema tribe. Did your opinion change as you read the box? Why or why not?
2. What image comes to mind when you think about this tribe. Write or draw the image evoked by reading this article description.
3. Choose a sentence in the box that was most interesting to you. Why was this the case?
4. What is the latipso?; Who are the witch doctors?; Who are the holy mouth men?; What is the shrine box?; Who are the Nacirema?

The story portrays people of a "tribe," described as fairly savage to both their own bodies and to each other. The box gently leads to the different institutions of the Nacirema society, each depicted with base and abhorrent rituals practiced. Often, when this story is discussed in the classroom, the Nacerima are judged harshly by the students. The twist to the story comes only after readers see a parallel (after much discourse) between U.S. society and the Nacirema. Of course, "Nacirema" is "American" spelled backwards. Did you realize this paradox by the end of the textbox?

*Source:* Miner, Horace, 1988, *Down to Earth Sociology*, The Free Press: New York, NY reproduced by permission of the author and the American Anthropological Association from *American Anthropologist*, 58:3 (1956), pp. 503–7.

This exercise is a strategy to help readers see their own egocentrisms and sociocentrisms. It is meant to break the inability to openly criticize our society as others would do from the outside. The suggestion by Paul to overcome sociocentrism to develop critical thinking is fundamentally important for scientists. The discipline requires that there is an objective reality and that one should think about one's own biases, both personally and societally, to overcome them. The scientist should make decisions about data based on his or her own opinions and not be influenced by the greater science community.

In this way, new insights and discoveries are made in science. Otherwise new ideas are foreclosed by mere repetitions of existing knowledge. The Nacirema story eludes most of us because we have a tough time of seeing ourselves and our culture from the outside in.

Scientists are often depicted as emotionless and valuing only the objective. The truth is that scientists are humans with feelings. Paul suggests *exploring thought and underlying feelings* about objective data. There is a real connection between the two that is necessary to recognize. For instance, a medical doctor may really like a patient personally and somehow overlook serious symptoms because he or she wishes that the patient is not gravely ill. At times, the medical community must practice removing themselves emotionally from a physical exam to evaluate objectively. This requires not only practice but an iron will.[9]

Affective strategies also involve an altering of confidence levels. On one hand, Paul suggests *developing intellectual humility* and on the other hand, *developing confidence and courage in reasoning skills*. Both suggestions seem to contradict but not really; each lead to the best conclusions based on objective truth. A humble person will not get caught up in the power of his or her own reasoning and is more likely to see all sides. However, the courage must also follow to make a reasonable decision and stick to it. If the tenets of this text are followed, the scientist can be confident that his or her strategies are reasonable and his or her ideas are solid. Confidence is gained by advance made with objectivity. Nonconformity and standing up to political or other social forces to uphold objectivity is its own reward. Paul's final affective strategy, to *develop intellectual good faith or integrity*, will prevail in the end. Honesty is the first rule in science and the center of integrity as discussed in other chapters. The ontological result will be the advancement of science.[10]

## Cognitive Strategies

Scientists deal with reality in simplified forms. Natural phenomena are actually complex but science requires models and oversimplifications to make sense of different happenings. That said, scientists must *avoid oversimplifications or generalizations* to make sure to study the complexity of the larger situation. Physiology of the liver can be taught and understood easily by listing the functions of the liver. This representation is then viewed by students as a discreet list of its physiology. In reality there is more going on than we know. There are complex interactions within the liver, more than a simple list given to students can realistically cover. The liver works with the blood to package fats, for example, to generate cholesterol for transport to cells. Each person is very different in ways we do not yet fully understand. Some are prone to develop high amounts of "bad" cholesterol which form plaques on artery walls and some are not. Why does the liver of one person eating bacon every day not package bad cholesterol while the healthy person suffers high fat levels? Genetics and liver physiology interactions are quite complex and perhaps beyond our knowledge. Many factors of liver function play a role in producing "good" and "bad" cholesterol levels. Science thinking often requires many levels of understanding, as seen in liver-fat processing.

Instead, Paul argues that *analogous situations should be compared to give insights to new contexts*. In this way, existing ideas can be explored (and expanded) or applied to new situations to further theories and pattern statements. For example, when presenting liver physiology as a subject to be taught, links to a lower fat diet or even effects of alcohol abuse on liver cell structure and function make comparisons that further thinking about content.

In order to *develop one's perspective by creating or exploring one's beliefs, arguments, or theories*, a few cognitive strategies should be in place. First, a clear statement of a thesis should be sought to ensure the investigation is well defined and understood. *Thesis statements* set forth the parameters of the argument and discussion in order to evaluate the given information. These should be clearly stated for a critical evaluation of the argument or topic. Second, *standards for an evaluation* should be developed to give a look into any investigation's results to determine if they are objective and clear. Third, the credibility of the *sources* for data should be thoroughly evaluated. Many sources do not fit the parameters of a study or are questionable . . . Does one take the barber's word that they need a haircut? Sources should be scrutinized for **validity** (how well it measures what it is supposed to measure) and **reliability** (how consistent the results are that they produce). Fourth, *analyzing or evaluating arguments, interpretations, beliefs, or theories* encompasses a set of strategies. Perhaps the most important aspect is to evaluate counterarguments to one's beliefs. The best critical thinkers are able to consider the points of the argument against their own. Also, it is important for the critical thinker not to simply overlook or dismiss the tenets of the counterargument. Objectivity requires that a counterargument is considered seriously until disproven. The ten quotes listed in **Box 11.2** are "golden oldies" that have been acquired over the years. They assist the critical thinker in evaluating and questioning another person's argument. They are also useful tools for success in a debate that requires argumentation skills. These strategies help to reach the higher stages of scientific argumentation discussed throughout the text.

Good critical thinking skills require reasoning *dialogically,* as discussed in another chapter. Thinking to oneself and speaking to others to get at as many perspectives as is possible are excellent strategies to critically reflect on scientific phenomena. Through this process, **dialectic thinking** is practiced, in which conflicting ideas are set against each other to determine the strengths and weaknesses of each. To accomplish this, the *Socratic method* of discussion is needed, which involves asking the right questions to get at those strengths and weaknesses. Any effective lecture requires an employment of the Socratic method to get students thinking as well as the lecturer.

---

**BOX 11.2   10 BEST SCIENTIFIC ARGUMENTATION COUNTERPOINTS (GOLDEN OLDIES)***

"THERE ARE MORE VARIABLES THAN CONSTANTS."

"NOTHING'S BLACK AND WHITE ABOUT THIS ISSUE."

"WHAT ARE YOUR SOURCES?"

"WHO'S FINDING YOUR SOURCES?"

"STATISTICS CAN BE DECEPTIVE."

"THAT'S A FAR STRETCH."

"THAT'S OVER THE TOP."

"YOUR ASSUMPTIONS ARE FLAWED."

"WHAT ARE YOUR ASSUMPTIONS?"

And finally the best of all:

"WHAT ARE YOUR PARAMETERS?"

*Use these for any scientific argument and your opposition will crumble!

---

These methods allow for another of Paul's strategies, to *distinguish relevant from irrelevant facts*. A thesis weakens as more irrelevant data obscure it. This requires strategies to think about one's own thinking, called *metacognition*. Paul states "one possible definition of critical thinking is the art of thinking about your thinking while you are thinking in order to make your thinking better: clearer, more accurate, and more fair."[11] It is clear that a use of certain vocabulary should be used during metacognitive thought: *conclude, assume, bias, relevance, evidence, justify,* and *consistent* are all terms useful for evaluating scientific claims.

Finally, the importance of *evaluating evidence and the alleged facts* in critical thinking is not to be understated. Paul suggests "not everything offered as evidence should be accepted. Evidence and factual claims should be scrutinized and evaluated. Evidence can be complete or incomplete, acceptable, questionable, or false."[12] Deep knowledge in the subject area, of course, advances the strongest possible evaluation of arguments and data supporting those arguments. Content knowledge is vital for critical thinking. The chapter on a modern synthesis treats an introduction to foundational knowledge within the branches of science.

No one is a superb critical thinker. We are all taken in by bias, confusion, and society at many turns, depending on the issue. The key is to practice critical thinking skills in the ways mentioned to develop into a more critical reasoner. Science requires this of its scientists and society requires this of a scientifically literate population.

## How to Think Critically in Our Society: Facebook Study

The reader is invited to engage in a treatise on how scientists think critically about the results of research reported by the media. This chapter describes the role of critical thinking and useful strategies in evaluating scientific research. Examples throughout the text expose how easy it is to make erroneous decisions based on media bias, inaccuracy, and poor methodology and foundational science. The reader will best become savvy to the pitfalls of misinterpreting and misusing science data by practicing critical thinking strategies in studying research.

To illustrate, consider the following report on the major U.S. news networks in October 2011. The "discovery" was made that "people who have more friends on Facebook have better developed regions of the brain." The reports were based on the abstract (summary) of a research study by the *Proceeding of the Royal Society B: Biological Science*, a journal reporting research in neuroscience. In it, 165 participants' brains were scanned using MRIs and each subject was asked to report the number of Facebook friends she or he had. Examine the abstract of the investigation in **Box 11.3** and use your strategies for critical thinking, research methodology, and mathematics to determine the relevance of the study.

## Media's Role in Critical Evaluation

More important than the study was the media's irresponsible reporting of the results. Immediately after the Facebook study was published, many news reports claimed that Facebook socializing helps people to develop portions of their brain to gain better skills in social functioning. Thus, the media concluded that Facebook helps people socialize to gain friends and popularity. In the media's articles, the conclusion is a simple relationship, without giving data on who funded the research or how many subjects were involved in the study. The types of measures used in studying the so-called "developed areas" of the brains under observation were also never addressed in the announcements. The media can be very useful in getting science information out to the public in an accessible manner but the obvious omissions mentioned lead one to question whether it is a good means of scientific transmission. Upon further investigation into the elements of the study, it was found that there were only 58 participants. The authors admit that the study may lack statistical power based on a small sample size. Additionally, on the small sample was composed of only college students. Thus, generalizability to the larger population is limited and media announcements were overblown and unwarranted.

What drives the media to make such claims to the public? A commonly cited explanation is that there is limited space or time available for news organizations to give all of the facts; that if a person is interested in knowing more, she or he can research the topic. Given the poor science literacy skills reported by international assessment measures, this assumption of public knowing is irresponsible. There must be more to the media's poor reporting of science. After all, sports reports and articles give far more data and have more time and space devoted to them than science research findings.

## BOX 11.3  FACEBOOK-FRIENDS STUDY

The increasing ubiquity of web-based social networking services is a striking feature of modern human society. The degree to which individuals participate in these networks varies substantially for reasons that are unclear. Here, we show a biological basis for such variability by demonstrating that quantitative variation in the number of friends an individual declares on a web-based social networking service reliably predicted grey matter density in the right superior temporal sulcus, left middle temporal gyrus and entorhinal cortex. Such regions have been previously implicated in social perception and associative memory, respectively. We further show that variability in the size of such online friendship networks was significantly correlated with the size of more intimate real-world social groups. However, the brain regions we identified were specifically associated with online social network size, whereas the grey matter density of the amygdala was correlated both with online and real-world social network sizes. Taken together, our findings demonstrate that the size of an individual's online social network is closely linked to focal brain structure implicated in social cognition.

Facebook Friends Questions

1.  What kind(s) of research study is this: Is it experimental, correlational, scientific modeling, or nonexperimental research? Why?
2.  What is the strength of the results based on the study's design? Is it powerful or weak? Explain.
3.  Does society play a role in the formation of the study and interpretation of its results?
4.  Are there any extraneous variables? How do they cast doubt on the conclusions made by the researchers?
5.  Do you think researchers explored their underlying feelings about the results? Were they objective in their reporting of the conclusions?
6.  Do you think the researchers were thinking independently when they designed the study? To what extent might they have been influenced by their own sociocentricity or egocentricity?
7.  Are the results presented in the abstract oversimplifications or generalizations?
8.  Are there valid counterarguments to the conclusions made by the researchers?
9.  In your metacognition, when you use the recommended terms, does any particular term help you to become more critical of the research presented?
10.  Describe the link between the thesis statement and the data. Is there a strong or weak link?
11.  What other evidence or research design could be given to strengthen the conclusions made by the researchers?

*Source:* Online social network size is reflected in human brain structure, R. Kanai, B. Bahrami, R. Roylance and G. Rees; Proc. R. Soc. B 2012 279, 1327-1334 first published online 19 October 2011.

Reporting the kinds of relationship interpretations in the above example of Facebook brain development may show both political and economic agendas to push technology—a kind of groupthink—to save the U.S. economy. If the public accepts that Facebook is good for it, then the part of our GDP dedicated to internet commerce and social media exposure online will strengthen. Or is it simple incompetence in reporting? Either way, the media acts as a help and hindrance to scientific thinking. In other chapters, we will explore the role of science knowing and science education filling the gap in science literacy to help the public become better consumers and users of science research.

# ■ KEY TERMS

affective reasoning strategies

cognitive reasoning strategies

critical thinking

dialectic thinking

egocentricity

metacognition

reliability

sociocentricity

Socratic method

thesis statement

validity

# ■ PROBLEMS

1. Which critical thinking strategy is most important in the development of a scientific hypothesis? An experimental design? Comparing scientific theories?

2. Which critical thinking strategy is missing in people who accept pseudoscience as valid science?

3. Why is critical thinking vital for scientific progress? Scientific literacy?

4. Which do you think is the *best* definition of critical thinking/reasoning/argumentation given in this text. That is, which most accurately reflects what scientists should know and be able to do?

5. Why do you value critical thinking in everyday life?

6. Discuss a problem you faced within the past week that required your implementation of critical thinking skills described in this chapter. Were you successful in addressing the issues? Why or why not?

7. Work in a team of friends to discuss the Facebook study. Gather their opinions on the study and see if another perspective emerges that you had not previously developed.

8. Research a topic of scientific interest to you. You may choose climate change, energy use, or criminal behavior to name a few examples. Be sure to use the strategies described in this chapter to evaluate the arguments and supporting data on the topic.

# ■ REFERENCES

1. Barnard, J. (1942). The lecture-demonstration vs. the problem-solving method of teaching a college science course. Science Education, 26, 121–132.

2. A Nation at Risk: The Imperative for Educational Reform. 1983. Report of American President Ronald Reagan's National Commission on Excellence in Education. Washington, D.C.: The Commission: [Supt. of Docs., U.S. G.P.O. distributor], 1983. v, 65 pp.; 23 cm.

3.  Daempfle, P. 2006. The effects of instructional approaches on the improvement of reasoning in introductory college biology: A quantitative review of research. *Bioscene: The Journal of College Biology Teaching*, 32(4): 22–32.

4.  Moore, B. N. and Parker, R. 1995. *Critical thinking*, 4th ed. (p. 4). Mountain View, CA: Mayfield Publishing Company.

5.  Lee, J. 2000. *The scientific endeavor: A primer on scientific principles and practice* (pp. 23–24). San Francisco: Addison Wesley Longman, Inc.

6.  Stegner, W. 1962. *Beyond the hundredth meridian: John Wesley Powell and the second opening of the American West,* Sentry Edition (pp. 214–215), Boston: Houghton Mifflin Company.

7.  Paul, R. W. 1992. Strategies: Thirty-five dimensions of critical thinking. In A. J. A. Binker (ed.) *Critical thinking: What every person needs to survive in a rapidly changing world* (pp. 391–445). Rohnert Park, CA: Foundation for Critical Thinking.

8.  Lee, J. 2000. *The scientific endeavor: A primer on scientific principles and practice* (pp. 23–24). San Francisco: Addison Wesley Longman, Inc.

9.  Paul, R. W. 1992. Strategies: Thirty-five dimensions of critical thinking. In A. J. A. Binker (ed.) *Critical thinking: What every person needs to survive in a rapidly changing world* (pp. 391–445). Rohnert Park, CA: Foundation for Critical Thinking.

10. Ibid.

11. Ibid, pp. 438–439.

12. Ibid, p. 444.

# CHAPTER 12

# A Modern Synthesis

## Earthquakes

A major earthquake occurs on the west coast of the United States. There is fire blowing through the cities and people scrambling to escape an incoming tsunami. The first thing one thinks of is the human tragedy associated with such a geologic event. The horrors of human damage come to mind on a personal level. Economic impacts are felt in an instant and rebuilding will take years. What took centuries to build up is destroyed in a moment. The possibility of a major environmental hazard, such as a radioactive leak from the local power plant, is feared.

Knowledge conquers such fear. In order to prepare for and minimize problems which arise in our natural world, science needs to be understood, drawing from all of the four science disciplines: physics, chemistry, biology, and geology. Earthquakes are a geologist's topic but we all share a stake in understanding the many facets of earthquakes. First, Earth's crust has large plates that move regularly since the formation of the Earth 4.5 billion years ago. These movements cause earthquakes at the edges of the plates. Geologists refer to this motion as *plate tectonics*, with Earth's plates depicted in **Figure 12.1**.

173

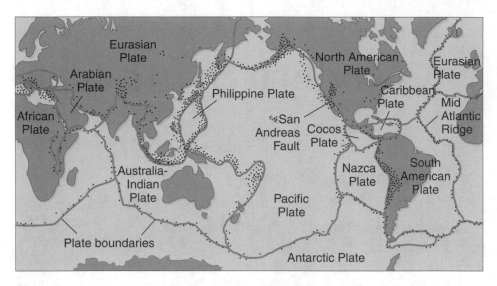

**Figure 12.1** The Major Tectonic Plates of the Earth. The dots represent earthquake locations. Earthquakes tend to occur near plate boundaries.

Plate movements are based on the laws of motion described in Newtonian physics. There are movements in an inner layer, the *mantle,* of the Earth. The mantle is thought to be composed of iron and nickel, in a molten or liquid form. The definite composition is still unknown. However, higher amounts of heat in the Earth's core, as opposed to the cooler temperatures near the Earth's surface, leads to temperature differences between the different regions. This difference drives movement of mantle fluid. As a result, plates on the surface either move away from each other or toward each other. Movement of the plates is termed *continental drift.* In fact, the rate of continental drift is about as fast as human fingernails grow. The land masses of Earth have moved significantly over the millions of years. **Figure 12.2** shows the direction and movement of the continents that have led to our modern day globe.

Movement of plates creates zones of *tectonic* activity, areas where the plates collide. Locations such as California, Japan, Indonesia, and Alaska are along the edges of these plates as shown in Figure 12.1. When the energy of the plates is released during the movements, resulting motion in the Earth above can be very severe causing much shaking and damage.

The energy from the layers within the Earth derives from the historical formation of the solar system. In fact, we are still cooling from the formation of the universe in what is termed the Big Bang. The Big Bang conceptualizes a time when all matter was created in a gigantic heated moment. All matter was formed from one giant release of energy. Matter was then scattered outward through space creating the *universe.* Astronomers believe the universe is still spreading outward today, as well. There is much evidence

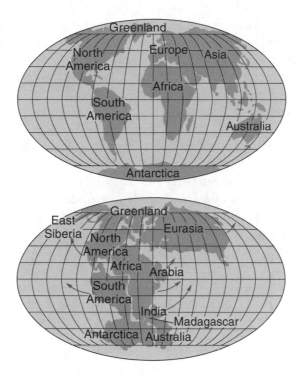

**Figure 12.2**    Continental Drift. (a) The Earth today and (b) about 200 million years ago. The motion of the continents in (b) is indicated by arrows.

for the *Big Bang Theory* but it is still a highly theoretical construct. The *Large Hadron Collider (LHC)* at the European Organization for Nuclear Research in Switzerland is attempting to recreate events during the Big Bang by bombarding particles close to the speed of light under supercooled conditions. A particle was formed from these collisions, called the *Higgs boson*. Its discovery gives insight into why matter has mass and if dark matter and microscopic black holes exist around us.

Geological disturbances result in threats to living species on the planet. Events such as earthquakes lead to major changes in the environment that sometimes are devastating for certain organisms. These organisms die off or change as part of the process Darwin described as *evolution*. Evolution is defined as a change in species over time. Evidence in the fossil records or from DNA taken from existing and former species helps determine the extent of evolution that took place. It is clear that life evolves along with changes in the planet. Almost 99% of all species that have ever lived are now extinct. The planet has changed drastically over its history, developing from a molten sphere to our living world today. In addition to earthquakes, several forms of catastrophic geological events have taken place. Volcanoes, meteor showers, ice ages, warming periods, flooding, and drying all have effects on living organisms on the Earth's crust. Observing and studying these events help us to explain how our Earth (both living and nonliving factors) came to be in its present form.

Modern species' *physiology* (the study of the way the body functions) changed through the generations to accommodate for changes in the environment. About 2.5 billion years ago, plant-like creatures produced large amounts of oxygen into the atmosphere. At some point, it became beneficial for other living things to be able to burn food using oxygen in small internal structures called **mitochondria**. Mitochondria are small structures within cells that make energy from food using oxygen. They are sometimes called the powerhouse of the cell. These organisms could operate much more efficiently than organisms that could not utilize oxygen. Thus, this advantage allowed for the many new species of what are termed *aerobic* organisms to develop and reproduce successfully. The ability to efficiently yield energy from food by using oxygen made modern *eukaryotes* very successful. Eukaryotes are living things that have a well-defined nucleus, or control center and have many structures within their cells. All living organisms today are eukaryotes except for the bacteria. Evolutionary scientists believe that eukaryotes developed from the more simple bacteria. Eukaryotes are more complex than bacteria. Bacteria are termed *prokaryotes,* and are more simple celled organisms lacking a true nucleus. The development of eukaryotes from prokaryotes is depicted in **Figure 12.3**.

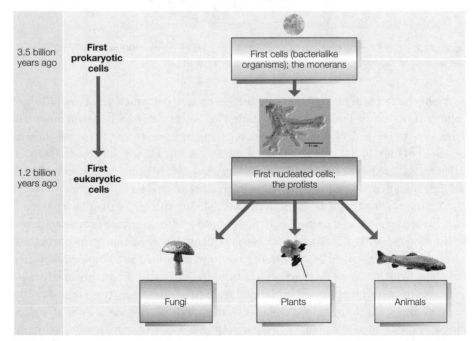

**Figure 12.3**   Evolution of the Prokaryotic and Eukaryotic Cells. This diagram shows the evolutionary history of life. Organisms fall into one of five major groupings, or kingdoms. The first form of life were the monerans, which are single celled prokaryotes. They gave rise to protists, single celled eukaryotes. Protists gave rise to plants, fungi, and animals.

As was illustrated by the earthquake example at the beginning of this chapter, for a phenomenon to be completely understood, it takes knowledge from all of the different branches of science. Full appreciation for what is happening in an earthquake draws from the physics that underlies the motion of the plates, the history of where the energy is coming from in the universe, the biological effects on both a medical and evolutionary level, and a vision of the chemical nature of the Earth's structure.

## Science as an Integrated Paradigm

Unlike mathematics, which is sequential and the depth of content follows a more clearly defined and measurable course, the sciences have many *interrelationships* between their branches. This contributes to a nonsequential nature for many of the science topics.[1]

Science should thus be seen as *integrated*, having all the areas draw from each other to understand phenomena. For example, teaching someone how to conduct DNA (deoxyribonucleic acid) testing without some knowledge of chemistry is useless. DNA is the hereditary material of life. *Nucleic acids*, subunits of genetic material in a cell, have base sequences that form codes that are fundamental to understanding DNA applications. There often needs to be background knowledge in all four science areas to understand even a lone concept in one.

With the explosion of information in science it is also no longer possible to deliver all of the information necessary to prepare science students for higher levels. Instead, readers should be shown how to organize that massive knowledge and see the underlying themes that emerge from the content. The earthquake example at the start of this chapter mirrors the kinds of explanations that should be given to answer real life questions in science. Instead of facing a problem of "covering all areas in the sciences," this chapter aids the reader in becoming able to more independently take control of his or her learning. The overview and synthesis shows the major themes and integration of the content to help readers organize the knowledge base behind science.

An overview of science themes is provided so that the reader can form greater conceptual understanding and place the information into meaningful networks. Science is seen as a whole, with separate parts contributing and interconnecting.

### Tracing Themes in the Four Sciences: Bringing the Sciences Together

In the introductory chapter of this book, science was described as a building up process, from mathematics to physics, chemistry, biology, and finally to geology. While it is true that one builds upon the other, each of the scientific disciplines also contributes specifically as a piece of a larger puzzle. I've heard many students of biology say, "I hate chemistry but I want to be a doctor," knowing that the student will face obstacles if they do not accept the importance of the integration.

There are a number of themes within the different science disciplines. As a first step, let's trace the movement of a *mole* of oxygen through our different subject areas.

Oxygen is fundamental to our existence and its understanding integrates many areas of science. To start, a mole of oxygen is $6.02 \times 10^{23}$ molecules of oxygen. That is a lot of molecules! Mathematics shows us the number: 602,000,000,000,000,000,000,000 molecules of oxygen. Chemistry gives us the conversion of this number to the number of molecules in a given volume of gas. A mole of oxygen gas takes up 22.4 liters of air at a normal/standard pressure and temperature. That is approximately 6 gallons of air. Using the *second law of thermodynamics,* which relates the motion of molecules to their energy levels, we can understand the motion of the gas related to different circumstances. According to the second law, oxygen molecules should flow into an organism through their random motion and tend toward increased randomness. In the case of a eukaryotic cell placed inside a container with one mole of pure oxygen, the oxygen molecules will flow away from the organized mole of oxygen to where they are needed, the cell's mitochondrion in a living cell.

Biologists can trace this oxygen and determine what happens to it after it enters the cell. A most impressive result is the uptake of oxygen as a final electron acceptor in the electron transport chain phase (a series of reactions in the mitochondrion) of *cellular respiration.* Cell respiration is the making of energy from food and oxygen that has been taken into a living system. Cell respiration requires a final step of converting oxygen to water, which elicits energy from food taken into the cell. This is the whole point of cell respiration. Without oxygen, a cell dies because it cannot obtain energy, which is needed for life to exist. A diagram of this process is shown in **Figure 12.4,** with an inner membrane of the mitochondria making energy in the form of adenosine triphosphate (ATP) as hydrogen and electrons travel through it to transform oxygen ($O_2$) into water ($H_2O$).

The *oxygen revolution* occurred, according to geologists, about 2.5 billion years ago. It came about due to the evolution of algae and primitive plants that carried out *photosynthesis* and yielded oxygen to the atmosphere, as mentioned earlier in the chapter. Photosynthesis is the making of food (glucose) from sunlight, carbon dioxide and water. This resulted in an availability of oxygen to the cells that could use it. The advantage of using oxygen led to an increase in the number of *aerobic cells.*

*Cellular respiration* is the basis for our modern human existence. We were able to yield fast and plentiful energy as a product of cell respiration. Therefore, this energy gave aerobic cells an advantage to live on and develop into all types of simple and higher-level organisms, such as humans.

Humans have done a great deal with oxygen, technologically. Oxygen is necessary for the internal combustion engine to function. Power from the engine comes from the burning of fossil fuels, such as oil and natural gas, to yield energy. The parallel with cell respiration is intriguing. Our cars, trains, jet planes, and home heating are based on applications of knowledge from principles of oxygen behavior and burning of food sources.

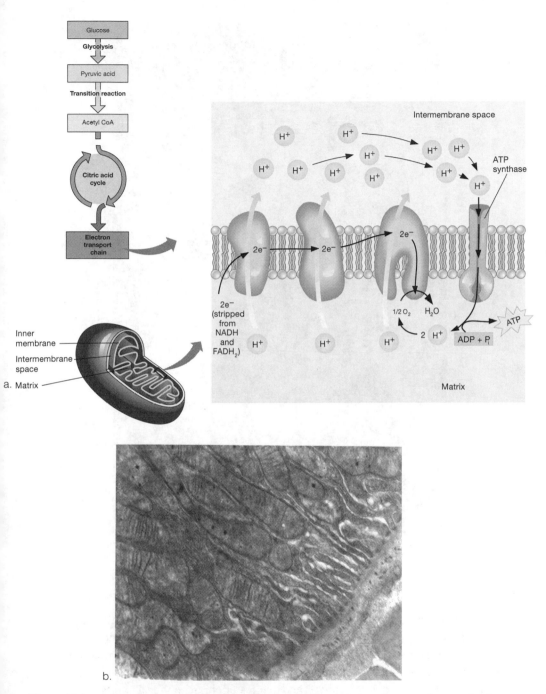

**Figure 12.4**    a. Cellular Respiration in Mitochondria b. Transmission electron micrograph of mouse mitochondria (380,000x).

© AbleStock

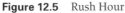

**Figure 12.5**    Rush Hour

Much of our economy is based upon the science of the internal combustion engine. Without this power source, our economy would halt fairly quickly. However, there are ecological problems with emissions from oxygen-based energy production as shown in **Figure 12.5**. Carbon emissions from fossil fuel combustion are implicated in global warming, pollution of the atmosphere, and deforestation. Geopolitical impacts require decision makers to have an understanding of oxygen and its applications in order to fully address societal problems.

## Science Literacy: Things Everyone Should Know in the Sciences

The theme of energy production, from its basic physical laws to its history and modern impacts, illustrates how science is integrated and how scientists need to work together to study phenomena. The oxygen example was placed in this chapter to emphasize that knowledge of science is vitally important in becoming a scientifically literate person. I often hear that science is only a process and as such no knowledge is needed—everything can be looked up. This is a fallacy. A working and effective base of knowledge is foundational to applying that information. The next section provides a synopsis of the salient principles of physics, chemistry, biology, and geology. It is not possible to give a comprehensive overview of science as a subject. All science cannot be simplified into a chapter or even a book. This chapter is only a launching point.

### Physics

Physics is the study of the properties, changes, and interactions of matter and energy. There are a few major themes of physics. First, *motion* is everywhere in the universe. Everything moves: Earth rotates, atoms vibrate, continents shift (plate tectonics), blood moves for oxygen exchange, people smell other people's scent molecules, a child's room gets sloppy within minutes (second law of thermodynamics, which states that all things tend toward disorder). The natural and random motion of molecules is termed Brownian motion, as shown in **Figure 12.6**.

Motion is caused by energy contained within a substance. More energy leads to faster motion. Motion is also caused by forces placed upon objects. Newton's laws of motion describe how a push or pull on an object causes movement. All objects resist motion: This is termed *inertia.*

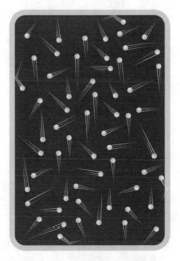

**Figure 12.6**   Brownian Motion

Newton's first law explains inertia, stating "an object at rest remains at rest and an object in motion remains in motion unless acted upon by a non-zero force."[2] This outside force must equal the mass of the object multiplied by its acceleration (how quickly it speeds up), $f = ma$ (where $f$ is the force, $m$ is mass and $a$ is acceleration). This is Newton's second law of motion. Thus, it takes a large force to accelerate a heavy vehicle. Alternatively, when comparing the stopping distance of two vehicles, the mass of the two cars determines which one will be easiest to stop, since the force needed to stop a large Mac truck is much more than that required to stop a small Prius. A larger mass has more forward energy and requires more frictional force to be applied to the brake pads to stop. Newton's third law of motion is called the law of action and reaction because every force has an equal but opposite force acting upon it, expressed as: $F = -F$. Consider when a gun recoils (jumps backwards) after it is fired. Larger guns like shotguns have greater recoil than small handguns. The larger the force placed upon the bullet (F), the stronger the jump backwards (–F). Even simply sitting on a chair while you read this book exerts a downward force that has an equal but opposite force pushing upward, as shown in **Figure 12.7**.

All chemical and biological structures obey these laws of motion. Consider the next example to apply Newton's laws. A duck stuck in an oil slick has a hard time staying upright or walking. The force of its feet doesn't have the opposite frictional force that normally allows the duck to more forward by pushing against the ground.

Newton's third law is limited by a loss of friction. *Friction* is any force that works opposite to motion. Oil in the slick prevents the grip of the duck's feet. The force of friction from the ground against the duck's foot is what the foot pushed on to give it

Force of chair on body

Force of body on chair

**Figure 12.7**   Newton's Third Law. When you sit, you exert a downward force on the chair, and the chair exerts an upward force on you. By Newton's third law, these forces are in equal but opposite direction. They do not cancel each other because they are exerted on different objects.

motion. Without this the duck cannot move. No force acts on the duck and due to inertia, it stops moving. Similarly, humans have fingerprints to make a roughened surface to give us more friction that is useful for picking things up. Without this friction it can be more difficult to climb or hold objects.

## Mechanical Physics

*Mechanical physics* deals with the motion of matter. Motion is based on *energy*. The amount of **kinetic energy** an object has determines how fast it moves as given by the equation Kinetic Energy $= \frac{1}{2}mv^2$ (where m is the mass of the object and v is its velocity or speed). If the energy of an object increases then so does the speed. Thus, for example, as the temperature of a smelly gas increases, the rate at which it moves around the room also increases. Some energy is stored, which is called *potential energy*. Potential energy in food is stored in bonds and is measured in calories. You know that as you eat foods high in energy either you burn it off in motion or it is accumulated as stored energy in fat molecules.

Energy occurs in *waves*. A wave is a wiggle or a vibration. It is a disturbance traveling from one place to another and always brings a certain amount of energy along with it. We started the chapter with earthquakes, which are energy waves in the Earth's crust and mantle. Heat, light, and even sound travel as waves. Different types of energy have different *wavelengths,* which is the distance between crests of a wave as shown in **Figure 12.8**. For example, green light has a slightly larger wavelength than blue light. Wavelength gives energy different characteristics including color seen by the eye.

Sound travels at a rate of 330 meters per second at 0°C in dry air. Sound is actually air molecules moving due to energy. When air hits human eardrums, the energy is concentrated by the ear's structure, as shown in **Figure 12.9**. Sound then hits the eardrum, which causes waves in the fluid of the inner ear or cochlea to move. Special cells in the *cochlea* (inner ear) have hairs that move when hit with the waves. The hair's movements lead to small chemicals flowing into the nerves along the cochlea. That flow is like an electric circuit, sending a current to the brain to let the person know that they heard something. Energy is demonstrated to be converted from one form (liquid wave) to another (electrical charge). Waves are present in so many aspects of the natural world from skipping stones on a pond to electrical activity in our brains giving us our personality. The presence of mechanical physics is everywhere.

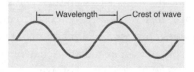

**Figure 12.8**   Wave

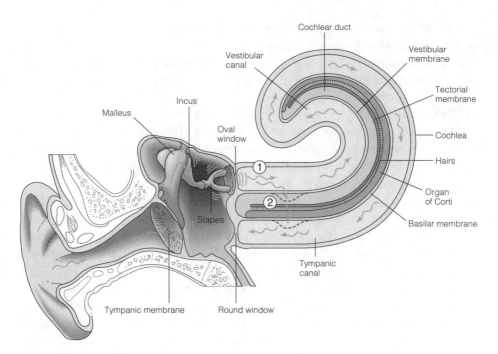

**Figure 12.9**    Ear Structure

## Nuclear Physics

Sometimes energy is released in large amounts from unstable substances. Such substances are termed *radioactive* because they release *radiation* as a form of energy. Substances that are radioactive can occur naturally. For example, radioactive carbon is made by photosynthesis in plants and radioactive potassium is found in all of our bones. They give off small, harmless amounts of energy. However, some decay and give off large amounts of dangerous energy: radium-226 and uranium-238 are examples of substances that give off a dangerous amount of radiation. They are naturally occurring and usually pose no harm because they are in such small quantities.

About 80% of the radiation exposure to humans is naturally present in the environment; it is due to natural geological sources. Sources for the other 20% are from cell phones, TVs, old fallout from 1960s nuclear testing still in the soil, nuclear power plant leaks, and other medical testing. In fact the radiation in one chest x-ray equals the natural exposure an airplane passenger receives on one round trip flight from New York to California, since your body is exposed to more galactic cosmic rays at higher altitudes. One trip does not pose a health hazard but repeated travel may. Flight attendants and pilots are exposed to more radiation than nuclear power plant workers, on average: 4.6 millisieverts (mSv) each year, compared to nuclear workers exposure of 3.6 mSv.[3] While diagnostic technicians and powerplant employees are afforded laws and protective measures from radiation, airline workers do not share these benefits. Some studies show a higher rate of cancer among flight attendants and pilots.

*Enrichment* is the process of concentrating natural sources of radiation to produce a variety of items: nuclear energy fuel, x-ray sources, smoke detectors, food safety treatment, cancer treatments, and atomic bombs. Radiation is used for our benefit in medicine, energy production, determining the age of organic matter, and perhaps war. A reasonable fear of radiation comes from the possibility of accidents with nuclear weapons or at energy plants, terrorists using radiation in dirty bombs to kill many people, and purposeful wartime weaponry. The social impacts of radioactivity are enormous.

The birth of nuclear physics was begun by the work of Marie Curie (1867–1934) who discovered its existence in natural geology and continued by several scientists. Albert Einstein (1874–1955), a theoretic physicist, never did experiments to split the atom, but expressed mathematically that such a process would produce large amounts of energy. In his famous equation $E = mc^2$, in which E stands for energy released equals m, or mass of a substance times c, the speed of light (which is large: $3.0 \times 10^8$ meters per second), this process was demonstrated. When applied in experimental physics, the discovery of nuclear fission and the atomic bomb was possible.

## Chemistry

The birth of chemistry comes from *atomic physics*. Chemistry is a study of matter based on the atom. Chemistry is the science dealing with the composition and properties of substances, and with the reactions by which substances are produced from or converted into other substances. The nature of *matter*—defined as anything that has mass and occupies space—has been a source of chemistry study as far back as human history goes. Gold, iron, lead, and copper are some of the pure substances studied by chemists. There are 92 naturally occurring pure substances, or *elements*. Elements cannot be broken down by ordinary chemical means. They are listed on a *periodic table* of elements generally ordered by increasing weight, as shown in simplified form in **Table 12.1**. *Atoms* are the smallest part of an element that cannot be changed by ordinary chemical means. Atomic physics developments gave rise to modern chemistry.

Most atoms come from stars long dead—from the *fusion* of original hydrogen atoms derived during the Big Bang. Fusion is the combining of atoms to form a new atom of a larger element that releases energy. As discussed earlier in Brownian motion, atoms constantly move, whether in the form of a solid, a liquid, or a gas. In addition to moving around, atoms also react with one another to form different substances. The atoms of the universe have been constantly combining and recombining since the beginning of time. Chemistry explains space, the universe, and Earth's origins. For example, lunar (moon) rocks are compositionally similar to Earth's mantle, indicating that the moon came from the Earth at some point in the past. This is called the *Giant Impact Theory,* which asserts that the moon came out of the Earth after a giant meteor hit the planet (**Figure 12.10**). Chemistry deals with these types

| Table 12.1 | A Simplified Periodic Table | | | | | | |
|---|---|---|---|---|---|---|---|
| I | II | III | IV | V | VI | VII | VIII |
| 1 ← Atomic number (Number of protons) H ← Atomic symbol Hydrogen 1.0 ← Mass number (Primarily determined by the mass of the protons and neutrons) | | | | | | | |
| 3 Li Lithium 7.0 | 4 Be Beryllium 9.0 | 5 B Boron 11.0 | 6 C Carbon 12.0 | 7 N Nitrogen 14.0 | 8 O Oxygen 16.0 | 9 F Fluorine 19.0 | 10 Ne Neon 20.2 |
| 11 Na Sodium 23.0 | 12 Mg Magnesium 24.3 | 13 Al Aluminum 27.0 | 14 Si Silicon 28.1 | 15 P Phosphorus 31.0 | 16 S Sulfur 32.1 | 17 Cl Chlorine 35.5 | 18 A Argon 40.0 |
| 19 K Potassium 39.1 | 20 Ca Calcium 40.1 | | | | | | |

Note that each element is represented by a one- or two-letter symbol. Elements are listed according to their atomic number (the number of protons). Their atomic mass is also shown.

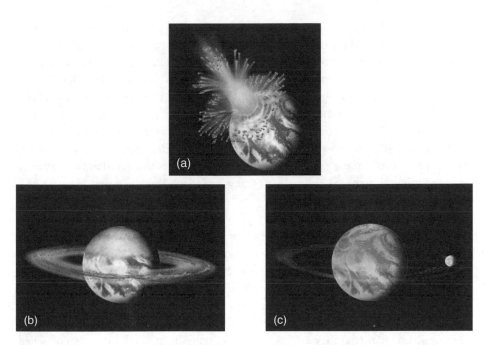

**Figure 12.10**    Giant Impact Theory. How the moon came to be: Young Earth was hit by a large object, perhaps as big as Mars (a). The two objects fused together but much material was thrown off and went into orbit around the new earth (b). Eventually the material coalesced into the Moon (c).

of submicroscopic details to understand larger questions about our world and the greater universe.

Humans themselves are a combination of atoms forming molecules, which in turn combine to form tissues, organs, and larger systems such as the skeleton. The sun is undergoing fusion, combining atoms to form larger ones, giving off the radiant energy we need to function on the Earth. That energy is cycled into our living systems and ultimately gives rise to life. Thus, energy is constantly being recycled into other forms.

Atoms also recycle. We are all made of the original stardust from the Big Bang. There are multitudes of atoms around us recycled from the past. They are particles unseen by humans and yet the essence of our existence. To illustrate, we breathe in 10 trillion particles per breath and do not see even one with the naked eye.[4] We should appreciate where atoms are coming from and going to—our past and our future.

## Inorganic Chemistry

Recall that atomic physics explores the structure of the atom and how it interacts with other atoms. *Inorganic chemistry* is the study of matter to reveal the behavior and structure of those atoms. This involves the study of so many different substances, from various salts, water, metals, and their combinations.

*Subatomic theory* is a branch of inorganic chemistry that deals with the three subatomic particles making up all atoms: *protons,* which are positively charged particles found in the center portion of an atom known as the nucleus; *neutrons* or neutral particles that are found in the nucleus as well; and negatively charged particles called *electrons,* which surround the nucleus and are attracted to the positively charged center. Much empty space exists between the electrons and the nucleus. In fact, Ernest Rutherford (1871–1937), in his famous *Gold Foil experiment,* demonstrated that the atom is almost 90% empty space. He shot particles through a solid but very thin sheet of gold foil and 90% of the particles passed through it. This showed that theoretically we are mostly empty space because living things are all made up of atoms. **Figure 12.11** shows the components of an atom and its nucleus.

Even smaller than subatomic levels, physicists are studying particles within the atom known as quarks. *Quarks* are parts of neutrons and protons (particles inside of atoms). Quarks are hypothesized to have even smaller parts, known as *superstrings,* which are levels of matter that make up matter at the smallest scale. Superstrings, or strings for short, are one-dimensional loops of matter that vibrate as an infinitely thin rubber band. This view of matter is known as *string theory* and has been mathematically but not experimentally shown. Technologically science is unable to "see" strings but mathematics shows that they are present in matter. Alternatively, it is also believed that dark matter, a force of energy and mass not directly detected so far, exists and makes up a large part of the universe. It is estimated that dark matter makes up 23% of the universe and its affiliated dark energy makes up 72%. Atoms and matter as we understand it compose only 4.6%. Dark matter and energy is indirectly detected but not seen or felt. Obviously, there is much more to learn about our universe than is currently understood. The organization of structure smaller than the subatomic level

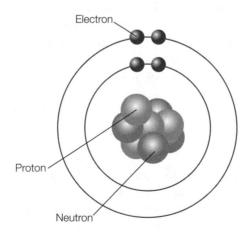

**Figure 12.11** The components of an atom and its nucleus

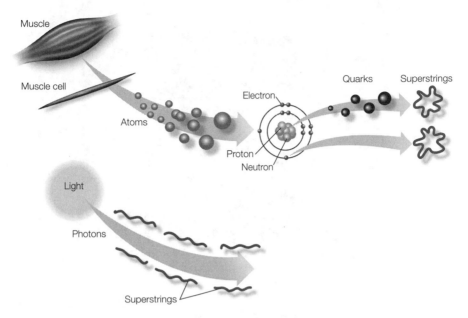

**Figure 12.12**    Superstrings

is given in **Figure 12.12**. It is possible that developments in these branches of physics will lead to major shifts in how we view the universe.

## Reactions

The electrons in the outermost shell around an atom are called *valence electrons.* They dictate the chemical activity of the atom. Whether a chemical combination or reaction will take place is dependent upon the energy of these valence electrons. *Chemical reactions* build up and break down substances by exchanging valence electrons. They form bonds, which are exchanges or sharings of electrons in the valences. Chemical reactions are symbolic representations of the different chemicals interfacing and forming a new product(s). Reactions are written using arrows between the reactants and products. **Figure 12.13** gives an example representing an exchange of valence electrons between two atoms. It results in the forming, breaking, and rearranging of bonds between the substances.

There are many types of chemical reactions, including: *synthesis,* which is the making of new substances from reactants; *decomposition,* or the breaking up of a larger substance into a smaller; and *replacement,* which is the exchange of one atom for another. A notable synthesis reaction is the process of photosynthesis carried out by plants, with energy to drive the reaction provided by the sun:

$$6CO_2 + 6H_2O \rightarrow C_6H_{12}O_6 + 6O_2$$

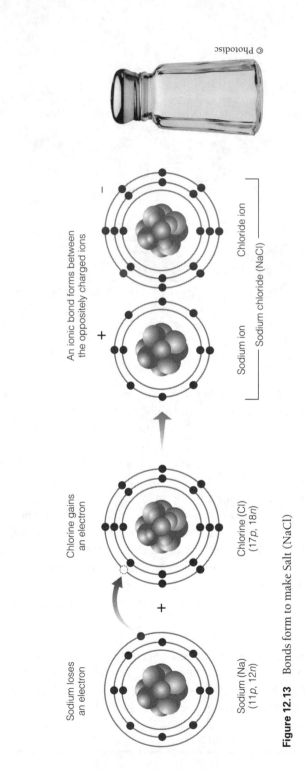

**Figure 12.13**   Bonds form to make Salt (NaCl)

This reaction provides the energy and oxygen for most life to function. The numbers in front of some of the molecules are there to show proportion, since it takes six molecules of each carbon dioxide and water to form one molecule of glucose and six molecules of oxygen gas. Chemists use double arrows when depicting chemical reactions that sometimes point in both directions. The type of reaction given by the double arrow is called a *reversible reaction*. One example of a reversible reaction is a set of chemical reactions by which the blood maintains stringent acid and base levels, called the carbonic acid-bicarbonate buffering system.

*Acids* dissociate in (often reversible) reactions producing lots of positively charged hydrogen ions. Conversely, *bases* react by removing hydrogen ions from solution. Acidity is measured on a *pH scale,* as shown in **Figure 12.14**, which ranges from 0 to 14, with 0 being the most acidic and 14 being the most basic. A pH of 7 is *neutral*; pure water is an example of a neutral substance. Many times the acidity of a substance has an important effect. Human blood requires a pH of between 7.35 and 7.45. Any deviation from this range is dangerous. *Respiratory acidosis* is a condition in which a person's blood acidity falls too low and the person may die if not treated.

Chemical reactions are really exchanges of energy. For example, just like energy to drive land masses during earthquakes, the burning of fossil fuels yields energy through decomposition, and your body uses the potential energy stored in your food to synthesize larger substances from simple matter. These examples demonstrate the nature of matter and energy exchanges during chemical reactions.

## Organic Chemistry

In general, living systems use energy to build up larger substances called compounds. Compounds form when two or more atoms chemically combine. A compound's basic unit is known as a molecule. Organic compounds are based on carbon as a backbone to the substance. Complex carbon-based chemicals make up the molecules of life. *Organic chemistry* is the study of carbon containing compounds.

Why carbon? Carbon forms large, long complex molecules because it has four valence electrons and forms four bonds with its neighboring atoms, as shown in **Figure 12.15**. This is a lot of bond potential. More bonds mean more relationships with other atoms and more complexity in those arrangements. In fact, there are 13 million organic chemicals vs. only 300,000 other inorganic (noncarbon) chemicals in existence.[5]

Simple examples include propane ($C_3H_8$), octane ($C_8H_{18}$), and enormous fat molecules called triglycerides. Reactive organic chemicals have different groups of atoms on them (e.g., $-NH_2$ and $-COOH$), which act chemically in a certain way to make each organic chemical behave uniquely. They provide a function for the substance and are called functional groups. This allows for a vast variety of uses for organic chemicals from penicillin and propane to aspirin and oil. In living systems, larger carbon-based substances called *macromolecules* carry out life's functions.

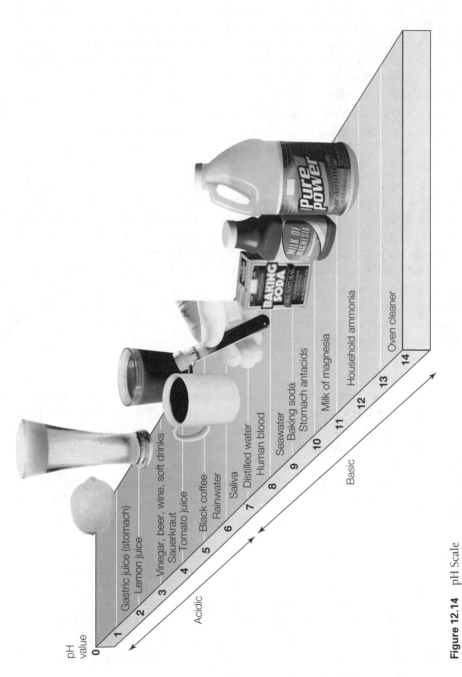

**Figure 12.14**   pH Scale

Lemon - © Photos.com; Beer - © AbleStock; Coffee - © Digital Stock; Vial of blood - © Comstock Images/Alamy Images

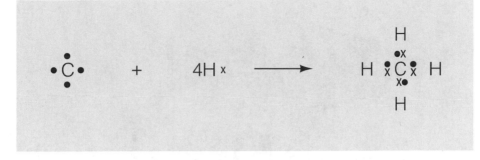

**Figure 12.15**   Why Carbon? Carbon forms four bonds with its neighbors.

There are four major groups of macromolecules in living organisms: proteins, lipids, carbohydrates, and nucleic acids. Each of these has specific functions to give the great creation of life. *Proteins* are a building block making up everything from hair, nails, and skin to hormones and enzymes that perform vital functions. *Carbohydrates* give instant energy to a living creature and *lipids* (fats) allow storage of energy in the longer term. *Nucleic acids* direct the functions of the living creature and the information stored in them is passed on from generation to generation.

## Biology

Let's review the organization of matter so far. Atoms organize to form molecules, which form macromolecules and life. **Figure 12.16** shows the organization of chemical substances into a living organism. Macromolecules organize in living systems in such a manner to produce life. Based on their chemistry, macromolecules combine to form *organelles,* which are structures that carry out specific functions in living systems.

These structures organize together inside a membrane or cell wall to form a cell. *Cells* are the basic functional unit of life. Cells that perform the same functions and have similar structure are called tissues, with muscles or nerves as examples. *Tissues* come together to form organs, which are specialized body parts that perform certain functions for a living system. *Organs* such as the kidney filter blood and the intestine, which absorbs foodstuffs. Organs working together comprise an *organ system.* The digestive system is made up of a few organs, the liver, intestines, stomach, and pancreas, which process food entering the body. Together the organ systems form a whole, living creature known as an organism.

The fundamental unit of an organism is the cell. A cell is like a city, with different parts working together as a single, whole entity. Some organelles found in cells include a *nucleus* or center that directs the cell's activities; a *Golgi body,* which modifies and purifies materials within the cell; and an *endoplasmic reticulum,* which acts as a subway to transport materials within the cell. A picture of the cell is given in **Figure 12.17,** with its accompanying description of organelle functions in Table **12.2.**

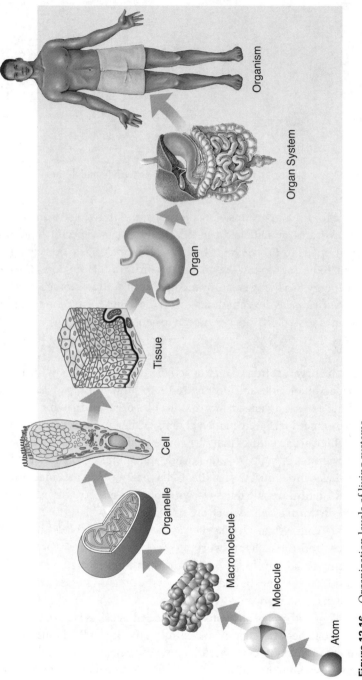

**Figure 12.16**   Organization levels of living systems

Adapted from Shier, D. N., Butler, J. L., and Lewis, R. Hole's Essentials of Human Anatomy & Physiology, Tenth edition. McGraw-Hill Higher Education, 2009.

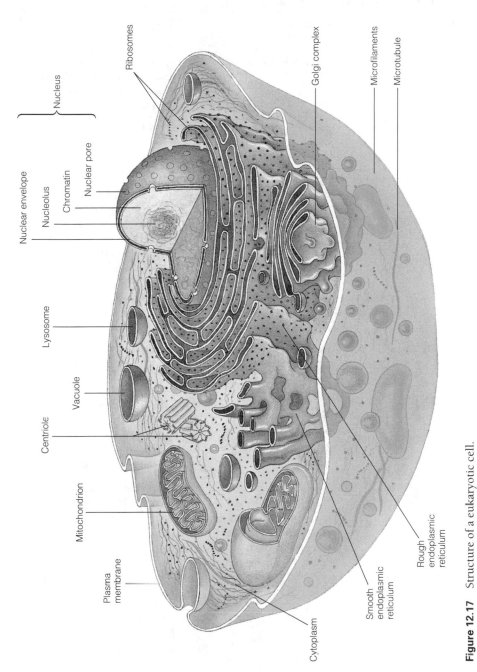

**Figure 12.17**  Structure of a eukaryotic cell.

| Table 12.2 | Overview of Cell Organelles | |
|---|---|---|
| Organelle | Structure | Function |
| Nucleus | Round or oval body; surrounded by nuclear envelope | Contains the genetic information necessary for control of cell structure and function; DNA contains hereditary information. |
| Nucleolus | Round or oval body in the nucleus consisting of DNA and RNA | Produces ribosomal RNA |
| Endoplasmic reticulum | Network of membranous tubules in the cytoplasm of the cell. Smooth endoplasmic reticulum contains no ribosomes. Rough endoplasmic reticulum is studded with ribosomes. | Smooth endoplasmic reticulum (SER) is involved in the production of phospholipids and has many different functions in different cells; round endoplasmic reticulum (RER) is the site of the synthesis of lysosomal enzymes and proteins for extracellular use. |
| Ribosomes | Small particles found in the cytoplasm; made of RNA and protein | Aid in the production of proteins on the RER and polysomes |
| Polysome | Molecule of mRNA bound to ribosomes | Site of protein synthesis |
| Golgi complex | Series of flattened sacs usually located near the nucleus | Sorts, chemically modifies, and packages proteins produced on the RER |
| Secretory vesicles | Membrane-bound vesicles containing proteins produced by the RER and repackaged by the Golgi complex; contain protein hormones or enzymes | Store protein hormones or enzymes in the cytoplasm awaiting a signal for release |

(*Continued*)

| Table 12.2 | Continued | |
|---|---|---|
| Organelle | Structure | Function |
| Food vacuole | Membrane-bound vesicle containing material engulfed by the cell | Stores ingested material and combines with lysosome |
| Lysosome | Round, membrane-bound structure containing digestive enzymes | Combines with food vacuoles and digests materials engulfed by cells |
| Mitochondria | Round, oval, or elongated structures with a double membrane. The inner membrane is thrown into folds. | Complete the breakdown of glucose, producing NADH and ATP |
| Cytoskeleton | Network of microtubules and microfilaments in the cell | Gives the cell internal support, helps transport molecules and some organelles inside the cell, and binds to enzymes of metabolic pathways |
| Cilia | Small projections of the cell membrane containing microtubules; found on a limited number of cells. | Propel materials along the surface of certain cells |
| Flagella | Large projections of the cell membrane containing microtubules; found in humans only on sperm cells. | Provide motive force for sperm cells |
| Centrioles | Small cylindrical bodies composed of microtubules arranged in nine sets of triplets; found in animal cells, not plants. | Help organize spindle apparatus necessary for cell division |

*Biology* is the study of life, from the cell to a community of organisms. There are a couple of major themes in biology. First, that life is *complex*. When viewing subcellular structures such as organelles and their relationships, such complication in function is profound. Second, that life has *order*. The feather of a bird or even the organized structure of the silicon dioxide shell of a marine diatom humbles even the best builder. Third, that life is very *diverse*. There are 1.5 million species of creatures on the planet. Half of them are insects! It is estimated that there are between 5 and 30 million species not yet discovered. Fourth, life *changes* or evolves over generations depending on its environment.

There are many branches of biology. A few include: *microbiology,* the study of organisms not seen with the naked eye (99% of all living things); *pathology,* the study of disease; *anatomy,* the study of structure; *physiology,* the study of function; *genetics,* the study of inheritance; and *ecology,* the study of organisms in their environment. Each of these studies is based on the same four principles of living systems described above.

*Macro-level biology* investigates how organisms interact within the environment, both with each other and with nature. This area of biology looks at behavior, effects of nonliving environmental factors (such as water availability) on living systems, and human impacts on the greater ecosystem (environment). Issues such as climate change, overpopulation, and endangered species are potential macro-biology issues. Studies of *food chains* and *food webs,* which describe the way energy flows within the environment, is included in macro-biology. An example of a food web is given in **Figure 12.18.**

Organisms always interact in ways to maximize their energy intake, using the nutrients of a particular area. One can actually mathematically predict how long a creature will stay in a "patch" of resources based on a couple of variables representative of the organism and its environment. In fact, weaker, more desperate amphibian (frog) males will hold onto females (the resource they are trying to obtain) during copulation (sex) to a point of drowning them. Mathematics modelling shows that weaker males will have a difficult time finding another mate so it "pays" for them to stay on top of the female as long as possible. It is a major cause of female amphibian death in North America! Being able to predict and modify behaviors in animals is called the study of *behavioral ecology.*

Regardless of the environmental factor studied, there is always order and balance in the operations of natural life systems. Imbalances or disturbances in nature, such as a tree falling in a forest or a catastrophic fire, always leads to nature striving for a return to normalcy in an orderly and predictable manner. In the example of the tree fall, plants will repopulate, first with smaller and less shade-tolerant ones, and later with larger, shading trees. This is how a stable, older tree community forms.

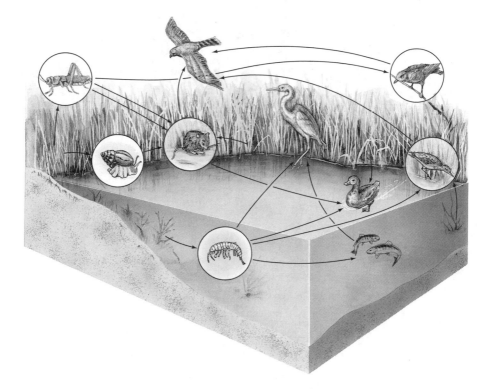

**Figure 12.18**   Food Web. Food chains are actually threads in a larger food web.

Conversely, *micro-level biology* studies the organism and its components. To illustrate, human health is an active area of study of workings of different parts of the body. *Diseases* are the focus. Disease is an imbalance in the internal controls of the organism. *Homeostasis* is the term used to describe this internal "steady state" as discussed in other chapters. Living systems control themselves in fast and slow ways. In humans, a *nervous system* uses a flow of charges to send messages to and from the brain in an effort to control muscles, bones, senses such as vision and hearing, and even personality. A diagram of the interconnectedness of the nervous system is given in **Figure 12.19**.

Slower means of control comes through the *endocrine system*. The endocrine system sends chemical messages, called hormones, through the body to communicate. For example, many brain activities such as thoughts or emotions are controlled by hormones. The living organism, much like the natural environment, strives to maintain order, balance, and normalcy.

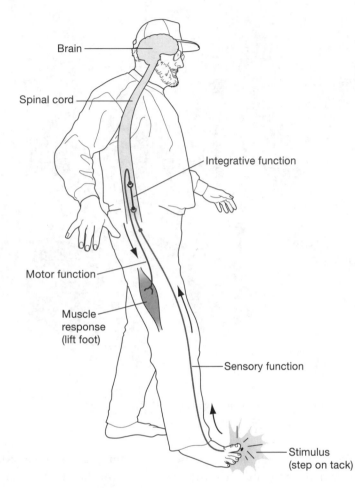

**Figure 12.19** Nervous System

There are obvious commonalities between branches of the sciences. Essentially, all study of biology involves understanding the chemical and physical nature of the components of life. Living organisms are linked to the energy of the Earth but are ordered in such a way that makes them the most unique creations known. Life is uniquely different from inanimate objects such as rocks. Nonetheless, as living beings, our link to the Earth's processes of energy transfer and matter recycling is undeniable.

## Geology

Earth's processes and its history are studied in the field of geology. Earth's history, since its formation ~4.5 billion years ago, is one of disturbance and change as discussed earlier. Ice ages, volcanoes, meteorite hits, and earthquakes are a few of nature's gifts. The Earth's energy comes from 13.7 billion years ago in the previously described *Big Bang*. All planetary motion and matter can be traced back to this time. Changes on our planet today are still driven by the sun, stars, and heat from the Earth's core (all remnants from the Big Bang).

Geologists divide the Earth's structure into three major partitions. The *crust* is the thin outer layer of the planet comprised of rigid continental and oceanic plates of the **lithosphere** and a weaker partially molten **asthenosphere** over which the continents (embedded in the lithosphere) drift like rafts slowly over long periods of time. Underlying the crust is a second massive layer, the *mantle,* composed of molten rock or *magma.* Convection currents in the deep mantle transfer heat from the innermost layer, the *core*, outward. The inner core is solid and the outer core is liquid, and crust movements are shown in **Figure 12.20.** These currents are responsible for the *plate tectonics,* which slowly move the rocky plates of Earth's lithospheric crust, driving many of the planet's geologic activities including volcanoes, earthquakes, and mountain formation.

There are about ten large rigid plates and several smaller ones that make up the lithosphere. Each moves as a discrete separate unit. Places where plates spread apart are **divergent junctions.** These are typified by earthquake activity and volcanism as the gap between the spreading plates fills with melted material from below and solidifies. Plates that separate in one place must come together in another. They may bump together in a head-on collision along a **convergent junction.** Mountain ranges are one result of these crumpled continental crust collisions while *subduction zones* are another. At a subduction zone one plate slips beneath the other at the site of a collision. As the edge of the lower plate is pushed deeper by the overriding weight of the upper plate it begins to melt under the heat and pressure, changing the composition of some rocks to form new ones and melting some of the lithospheric material back to magma (some of which makes its way back to the surface through volcanic vents). In addition to collision and separation of plates there are regions where plates simply slide along one another at a *transform fault*. Collisions of land masses are shown in **Figure 12.21.**

Earth's lithosphere is primarily comprised of rock. Rocks are classified into three general types: **igneous**, those formed directly from magma such as granites; **sedimentary**, those formed via the process of weathering and particles being compressed as with sandstone, shale, and limestone; and **metamorphic,** rocks of other types exposed to either extremely high pressures or temperatures that alter them significantly like those found deep in a subduction zone.

Rocks and their layers in the crust can be studied to retrace a record of geological history. Geologists study those rocks that were preserved from destruction through time knowing that layers are formed by deposits of sediment from air and water. Since

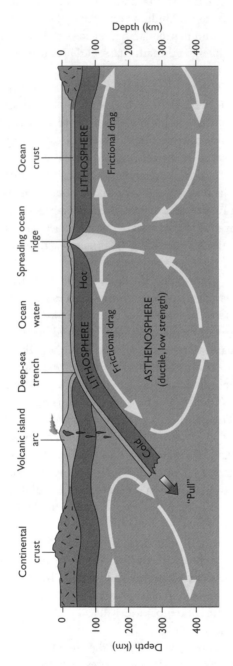

**Figure 12.20** Driving mechanisms for plate motions.

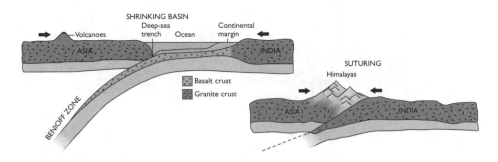

**Figure 12.21**    Collisions of continents

a sedimentary layer cannot be deposited beneath a previously deposited bed, naturally newer layers are always added on top (barring any deformation or overturning at a later time). These layers or *stratifications* and the time sequence under which they were laid allow a stratigraphic time scale to be constructed, which can then be used to analyze events that took place throughout geologic time on Earth. This allows geologists to recreate a history of the planet through geologic eras noting its conditions and life forms using fossils and dating techniques as well as the ability to study the rock record to trace lithospheric movements and events through time.

Geology is generally divided into two broad areas, historical geology and physical geology. *Historical geology* involves the evolution and origin of the Earth, its continents, oceans, atmosphere, and life. *Physical geology* examines processes operating within the earth and on its surface in addition to earth materials including rocks and minerals. Specifically, geology is divided into numerous branches including but not limited to: *petrology* and *mineralogy*, studies of rocks and minerals; *stratigraphy*, study of the rock layers in the Earth's crust; *seismology*, study of earthquakes; *paleontology*, study of fossils and ancient life; *astronomy*, study of the planets, solar system, and universe; and *hydrology*, *oceanography*, and *climatology*, studies of Earth's water and the atmosphere and their cycles. Clearly, the geosciences are very much grounded on basic chemistry and physics principles. Movements within the Earth, its crust, the solar system, and the universe are governed by the laws of physics. Cycling of chemicals between the regions of the Earth, termed *biogeochemical cycling*, allows nitrogen, carbon, water, and phosphorous to move through the earth, its biota (life), water, and atmosphere. This relationship is shown in **Figure 12.22**. Recycled chemicals are used in living systems, moved through water cycles and atmospheric processes, and deposited to the Earth for uptake again. The chemicals follow physical laws of motion and chemical reactivity. All of this interconnectivity demonstrates the very integrated nature of geology with other fields of science.

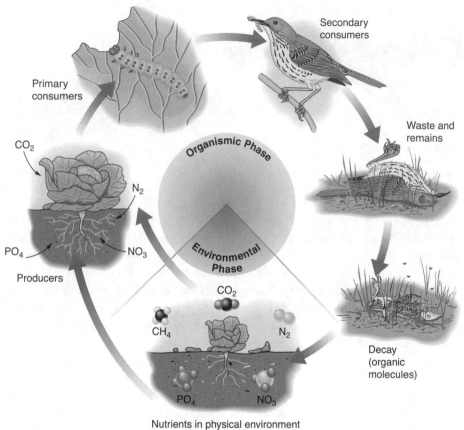

**Figure 12.22**   Nutrient Cycles. Chemicals are transferred between organisms and their environment. $CO_2$ = carbon dioxide, $PO_4$ = phosphate, $N_2$ = nitrogen, $NO_3$ = nitrate, $CO_4$ = methane.

## Teaching Science as an Integrated Subject

The vertical nature of science (that one builds upon the other) is a weighty barrier to successful understanding of content.[6] Since science curricula tend to be vertically structured, a certain level of content knowledge for incoming college students is necessary for success in college science courses.[7] Content learning expectations for sciences should thus "be considered intrinsically multidisciplinary. Student learning is enhanced when we are able to help students see the relationships among the sciences, and between science and mathematics, the humanities, social sciences, and the arts."[8]

Rather than focusing on amount of content necessary for learning, the *organization* of that content around themes, issues, or projects can enrich the students' view that the sciences are not separable from other areas of study and can be reasoned about in a more holistic way. Although there is no universal teaching strategy to accomplish this, research shows that some general principles are effective: (1) Teach *scientific ways of thinking* to explore all areas of the sciences; (2) *actively* involve students in their own learning; (3) help students to develop a conceptual framework to enhance *collateral* learning about other science areas; (4) promote *discussion* and group activities around *integrated* themes; (5) apply concepts learned to new, even *nonscientific situations* outside of the course; (6) develop *experimental* and data analysis skills to base science on *mathematics;* and (7) *test* important laws and rules of the different sciences.[9]

A part of scientific literacy is the ability to acquire a knowledge base including a certain level of understanding of content material in the science areas, defined by various national standards.[10] A foundation for understanding was provided by this chapter, with knowledge areas from national and state standards, addressed almost entirely in its presentation.

## College Science Standards

Developing a reader's understanding of college science content and perspective to match the academic demands of undergraduate faculty was a goal of this chapter. The Benchmarks for Science literacy (1993) and the National Science Education Standards (1996) nationally define the content, instructional, and assessment strategies appropriate for the K-12 science curriculum. However, no such set of guidelines have been effectively implemented or defined across the undergraduate science teaching fields.No Child Left Behind is a K-16 mandate, meaning that college reform is also required under the law. This chapter uses the nationally defined standards to present material in such a way to prepare students for an entering major in one of the sciences.

While the National Science Teachers Association (NSTA) apply the NSES standards, meant for K-12, to college science, few higher education science faculty are aware of its existence. In the College Pathways series presented by NSTA, all six science standards—teaching, professional, development, assessment, science content, science education programs, and science education systems—are applied to college science education. This text mirrors the content and presentation advocated by College Pathways. College Pathways uses vignettes and cases to show models of teaching for college science courses, much in the same way this text shows science principles.[11] The base information in this chapter is a key component in developing one's scientific literacy.

# ■ KEY TERMS

acid
aerobic
asthenosphere
atom
atomic physics
base
Big Bang theory
biogeochemical cycling
carbohydrate
cell
cellular respiration
cochlea
compound
continental drift
convergent junctions
core
divergent junctions
electron
element
enrichment
eukaryotes
evolution
friction
fusion
Giant Impact theory
Gold Foil experiment
Higgs boson
historical geology
igneous rock
inertia
inorganic chemistry
kinetic energy
Large Hadron Collider (LHC)
lipid
lithosphere
macro-level biology

macromolecule
mantle
matter
mechanical physics
metamorphic rock
micro-level biology
mitochondria
mole
molecule
neutron
nuclear physics
nucleic acids
organ
organ system
organic chemistry
organelles
oxygen revolution
periodic table
pH scale
photosynthesis
physical geology
physiology
plate tectonics
potential energy
prokaryotes
protein
proton
quarks
radiation
respiratory acidosis
second law of
    thermodynamics
sedimentary rock
strings
subatomic theory
superstrings

tissue                                          wave
transform fault                                 wavelength
valence electrons

# ■ PROBLEMS

1.  What is the distance of a thunderstorm when you note a two second delay between lightning and sound?
2.  Based on the wave theory of motion, is there sound in outer space? Why or why not?
3.  Which has a higher kinetic energy, 1 kg of gold at 40°C or 3 kg at 10°C?
4.  What causes the Earth's plates to drift?
5.  What are the major themes in the study of life? Which do you think is most interesting? Why?
6.  Group project. Select a team of four students. Choose a topic of interest. Make a concept map relating all of the science areas to that topic using the terms in this chapter. (For example, if you would choose genetic testing, areas from DNA chemistry, a look at DNA found in the Earth's crust, diffraction patterns of physics, and coiling in the nucleus of the cell might be related areas.)
7.  Build a collage of ideas centering on a topic you choose in one of the sciences. Relate, using pictures in the collage, the different branches of the sciences to the selected topic. Use a standard poster board and fill the collage with magazine or newspaper articles.
8.  How does Newtonian physics differ from nuclear physics? How did atomic physics give "birth" to both nuclear physics and chemistry?
9.  How does organic chemistry differ from inorganic chemistry? Name a compound or element within each of these branches.
10. List the three classifications of rocks. Discuss how each are formed.
11. Discuss how the following terms are related. Give one example on how they are related.
    a.  Geology and Physics
    b.  Chemistry and Biology
    c.  Geology and Biology
    d.  Art and Physics
    e.  Music and Chemistry
    f.  History and Biology
    g.  Philosophy and Geology

## ■ REFERENCES

1.  California Higher Educational System. 1984. *Statement on preparation in natural science expected of entering college freshmen.* California Community Colleges, Sacramento; California State University, Sacramento; Academic Senate, Sacramento. ERIC Document Reproduction Service No. ED242375 Available at www.eric.ed.gov.

2.  Hewitt, P., Lyons, S., Suchocki, J., and Yeh, J. 2007. *Conceptual integrated science* (p. 36). Boston: Pearson Addison Wesley.

3.  Ibid. pp. 195–196.

4.  Ibid. p. 663.

5.  Ibid. pp. 294–301.

6.  Westheimer, F. 1988. Education of the next generation of scientists. *Chemical & Engineering News* 66: 32–38.

7.  California Higher Educational System. 1984. *Statement on preparation in natural science expected of entering college freshmen.* California Community Colleges, Sacramento; California State University, Sacramento; Academic Senate, Sacramento. ERIC Document Reproduction Service No. ED242375. Available at www.eric.ed.gov.

8.  Ibid.

9.  American Association for the Advancement of Science. 1990. The liberal art of science. *Science* 1 June 1990: 1137.

10. National Research Council. 1996. *National science education standards.* Washington, DC: National Academy of Sciences Press.

11. College Pathways to the Science Education Standards. 2001. Edited by Eleanor D. Siebert & William J. McIntosh. Arlington, VA: NSTA Press.

# CHAPTER 13

# Science Education: The Need for Good People in Science

*"Choose a job you love, and you will never have to work a day in your life"*
(Confucius)

Getting people to love science and enter into a career in one of its branches is a goal of this chapter. Too often, students are turned off from the many exciting professional possibilities that science careers offer. Usually, experiences within the science classroom contribute to student dissatisfaction with a science major. This chapter explores the many factors that may lead to college students leaving a science major field of study. Research is presented in the hopes that science education will attract more qualified talent into the field. The chapter should also increase student awareness about the nature of science classrooms and how to improve a student's coping strategies to navigate through science programs today.

# Introduction

The overall goal of this book is to improve science literacy to help bring readers into the field of science. Too few college students are recruited and retained in science programs to meet the nation's future needs. The National Research Council indicates that first-year student interest in choosing college science, technology, engineering, and mathematics (STEM) majors has been in an alarming decades-long decline.[1]

More specifically, first-year student interest in overall undergraduate science majors declined from 11.5% in 1966 to 5.8% in 1988. The trend continues. The most dramatic declines were seen in mathematics (from 4.6% to 0.6%) and the physical sciences (3.3% to 1.5%).[2] Attrition remains a major concern for any science program. *Attrition* is defined as the drop-out rate from a field or major area of study.

A study conducted by UCLA's Higher Education Research Institute found that STEM students take longer to complete their degrees than non-STEM students. The study tracked thousands of students who entered college for the first time in 2004. Unfortunately, first-year college student STEM attrition was the highest of all college majors. The Center for Data Exchange and Analysis tracked students who entered STEM bachelor's degree programs in the 1993–1994 academic years. The study found that only 38% of these students earned a STEM bachelor's degree within six years.[3]

Why should a reader be interested in science education attrition? The prosperity and safety of the world depends on human talent entering into the sciences. As discussed throughout the text, economic, military, philosophical, and practical progress is made by human scientific thinking. Without human talent attracted into science, the nation and world weaken.

# The High School/College Science Divide

U.S. census data also show that potential majors in science, technology, engineering, and mathematics (STEM) are lost particularly in the transition from high school to college by freshmen switching from STEM majors to non-STEM majors.[4] A student loss rate of 40% occurs at this juncture (between high school and the first year of college) from STEM majors on the whole, with losses ranging from 50% in the biological sciences to 20% in the physical sciences and mathematics.[5]

The greatest drop-out rates for students was found among high school graduates who withdrew their decisions to enter a STEM major at or before enrollment in college.[6] However, during college, the highest risk of STEM switching (35%) occurred at the end of the first year.[7] As student time in college increases, risk of attrition declines, with some studies reporting a loss between sophomore and junior year of only 2% and from the start of the junior year to graduation, 0.8%. Interestingly, however, very few students transfer into STEM majors after college enrollment and there is always a net loss.[8] Kenneth Green, an educational researcher in STEM thus pointed out that "not only do the sciences have the highest defection rates of any undergraduate major, they also have the lowest rates of recruitment from any other major."[9]

Career entry in STEM areas has obviously been affected by these attrition rates, whereby both the health professions and engineering areas lost over half of their entrants in the past two decades (53% and 51%, respectively).[10] Particularly, a shortage in the supply of science teachers in various geographic areas was noted by the National Research Council. Considering the rise in college enrollment, the reported declines of 60% in the number of students preparing to teach science is alarming. Although a variety of factors have purportedly contributed to this situation of STEM teacher scarcity, clearly the high attrition rates for STEM college majors have played a role.[11]

The ramifications of this decrease in STEM enrollment are evidenced in the studies showing a declining scientific literacy of the population as a whole. Science literacy is the ability to know and be able to do science in the ways described in the definition of science elsewhere in this text. This change has produced a nation that has "simultaneously and paradoxically both the best scientists and the most scientifically illiterate young people . . ." according to Goodstein.[12]

Declining scientific literacy, combined with the decreases in STEM enrollment, has resulted in reduced numbers of qualified individuals available for not only science teaching but also research development, a driving force in the progress of science. Public concern has thus been expressed regarding the international competitiveness of the U.S. in the science and technology-dependent sectors of the economy.[13]

Because serious deficiencies in STEM education have been recognized by both the public and the scientific community, attempts should be made to reduce attrition and improve achievement among college STEM majors.[14] Thus, the major purpose of this chapter is to explore the empirical evidence explaining the increasing rates of attrition in STEM areas. Science teaching, cultural components, and the hope is to ameliorate any deleterious effects by the current educational system on the future of scientific development. Ways to improve science, mathematics, and engineering education for all students is the goal of this chapter. Socioeconomic effects of STEM deficiencies on U.S. competitiveness will be explored in another chapter.

## The State of Science Teaching

University faculty have traditionally explained undergraduate attrition from STEM majors as appropriate, claiming that the unprepared and lazy are weeded out. This may well be true, but what about the qualified students who leave? Many students are excellent scholars but are turned off by aspects of the science classroom and turned on by other majors instead. Additionally, research shows that many qualified candidates for a STEM career are weeded out in academic programs who might otherwise do quite well on the job.[15] A variety of studies have been conducted in order to tease out the variables that contribute to a student's decision to switch from a STEM major: Loftin in 1993, Razali and Yager in 1994, Strenta, Elliott, Adair, Matier, and Scott in 1994, Seymour and Hewitt, 1997, Daempfle in 2004, and McShannon in 2001.[16]

## Instruction and the "Chilly Climate Hypothesis"

The overall aim of an extensive study by Seymour and Hewitt was to establish the importance of factors influencing decisions of high achieving college science majors to switch from a STEM major to a nonscience-based discipline. Seymour and Hewitt conducted an ethnographic study over a 3-year period (1990–93) with 335 students in STEM majors drawn from seven university campuses containing a high proportion of these majors. Most of the data were gathered by personal interview. An additional 125 students participated in focus groups on six other campuses.[17]

In order to hold the variable of academic ability constant in explaining attrition, Seymour and Hewitt (1994) studied only students who were considered well prepared for college STEM majors, having math SAT scores above 649. Half of the students included those switching from STEM majors and the other half included the *persisters* (those remaining in the STEM majors).[18]

Surprisingly, Seymour and Hewitt reported the same set of concerns in both groups. Complaints were so serious that they led qualified students to consider switching from their STEM major and fomented strong discontent among the persister group. A set of 23 concerns was shared by switchers and nonswitchers alike. Nine out of ten switchers (90%) and three out of four persisters (75%) described the *quality of science instruction as poor overall*. Concerns about teaching effectiveness, testing styles, and curriculum (limited courses offered) in STEM majors appeared in all issues raised on the list. The results showed that students strongly believed that STEM faculty did not like to teach, did not value teaching as a professional activity, and valued their research above teaching.[19]

When asked to compare STEM courses with non-STEM courses, students expressed strong contrasts: coldness vs. warmth, elitism vs. democracy, aloofness vs. openness, and rejection vs. support. The most common words used by first-year students to describe their personal encounters with STEM faculty were: *unapproachable, cold, unavailable, aloof, indifferent, and intimidating*. Students further elaborated, describing the coldness of a STEM classroom as based on sarcasm and ridicule by faculty.[20]

These practices, rarely found in non-STEM courses, are described by students as discouraging voluntary student participation and creating an atmosphere of intimidation. Switchers cited these issues as a main cause of their decision to leave their STEM major. Attribution of this type of classroom structure to attrition is termed the *chilly climate hypothesis* in the literature and is hyperbolically depicted in **Figure 13.1**. The snake is the hostile science atmosphere and the student the victim.

In studies, criticism by students also focused on a lack of discussion in the college classroom, with only one-way lectures. Interestingly, students valued their high school

© Hemera/Thinkstock

**Figure 13.1**   Predator Control

experiences, which they described as containing much dialogue, over their college lecture courses in STEM areas. Poor preparation for classes by instructors, a focus on rote memory as a class goal, and faculty reading directly from textbooks were described by students as factors contributing to poor STEM instruction.[21]

A common explanation given by science faculty for the high attrition rates in STEM is poor high school student preparation. On the other side, students in a variety of studies value their high school experiences in science and math over college STEM courses, citing more interesting and better instruction. This does not negate the opinions of college faculty, but it indicates a difference in expectations of high school and college faculty for their students and a favoring of the high school situation by students. Maybe high school is simply easier or is there real differences in STEM approaches at these levels? Analysis of reasons for preference for high school instruction was limited in most studies to student interviews expressing displeasure with the chilly climate of college STEM vs. the nurturing environment of their high schools.[22]

Could there be an intrinsic difference due to content factors between science and nonscience areas? In other words, are nonscience subjects easier? Researchers Strenta, Seymour, and Hewitt in 1994 and Gainen in 1995 published research in which they asked students to elaborate on their reasons for leaving or their general unhappiness in STEM areas. Results showed that students were generally interested in the sciences but were "turned off" by the structure and climate of the classroom.[23]

Complaints about language problems with foreign assistants, large class sizes, and poor high school instruction are not helpful but the main issues students cited were poor classroom teaching: an intimidating classroom climate, poor quality of undergraduate science teaching (particularly dull lecturing), and a general lack of nurture for the student were cited most frequently. Studies showed that faculty may be able to reduce STEM student loss rates by improving classroom climate, changing from a competitive classroom structure by incorporating cooperative learning strategies to develop peer support.[24]

It is important for readers who may be considering science not to be dismayed by these studies but to become more aware of how to make STEM education better and how to avoid pitfalls that the current system may introduce. This chapter and its background research attempts to send a message to improve science teaching and learning to attract and retain more student talent.

## Personal Contact with Faculty

Many students cited a college faculty focus on research instead of teaching as a problem. However, student attitudes about faculty preoccupation with research and poor relations with students changed when students were allowed to participate in that research. The few students who had experienced this *faculty-student research* valued the open relationship with faculty in a research situation compared with the faculty's apparent indifference to them in a teaching context.[25] Other studies show mentoring experiences in undergraduate STEM areas are strong predictors of persistence. Additional factors that contributed significantly to the retention of STEM students included regular personal contact with a particular instructor who took interest in them, departmental gatherings, and small group learning and discussion.[26] Students entering a STEM career should seek to gain a research relationship with a faculty member early in their major. Through going to office hours, forming study groups, participating in department events and researching a professor's research, STEM majors can avoid the pitfall of getting lost among the crowd. It takes energy and focus but readers should be aware that most faculty want the interactions with their students—it is the key to forwarding their own thinking, with new student ideas.

Unfortunately, a frequent complaint among STEM students in several studies was inadequate personal contact with faculty during advisement. Many students traced their retention problems to improper faculty advice. Over one-third of both switchers and nonswitchers felt that they were not made aware of the length of study (e.g., more than 4 years) required of STEM majors and of the financial commitments needed to complete a program.[27]

## Women and Underrepresented Minority Groups in Science

There is a disproportionate rate of graduation between White and Asian American and *Underrepresented Minority URM groups* (which includes Latino, African American, and Native American). About 23% of URM STEM students earned a STEM bachelor's degree within six years as opposed to 41% of White and Asian American STEM students who earned a STEM bachelor's degree within six years. In 2004, four year completion rates for first-year STEM students were: White and Asian American groups had completion rates of 24.5% and 32.4%, respectively; Latino, Black, and Native American had completion rates of 15.9%, 13.2%, and 14.0%, respectively.[28]

Engineering and physics remain the most exclusively male-dominated STEM disciplines. In an analysis of national university enrollment and graduation data, minority women disproportionately fail to be recruited and retained in engineering and physics, with their first-year enrollments declining in the 1990s. The study blames a feeling of student isolation within the engineering community. More recent studies of engineering students found that minority women were more successful in nontraditional learning patterns (e.g., cooperative learning styles) than their cohorts.[29]

This indicates that retention among underrepresented minority groups in STEM areas would be enhanced by changes in instructional styles that include multiple learning style approaches and a departure from the competitive "chilly climate." The largest study to date, funded by the Sloan foundation and for which I am an active reviewer, is set to explore the nature of attrition in engineering. The study spans across a large segment of engineering programs and will publish its results in 2013. An important goal for the future of science is to retain women and underrepresented groups to add talent and expertise.

The overall numbers for attrition among URM and women are not pleasing, however. With low rates of retention, student interest in these groups is less able to materialize into STEM careers. Overall, the low rates of these students reaching their goals within a reasonable time frame indicate systemic problems in science education. What factors are contributing to such students leaving the STEM major?

Are there aspects of the STEM learning culture which contribute to these losses? It would be prudent to explore these elements and to develop institutional and curricular changes with these groups in mind. Attrition from STEM fields is a national issue and remains a major concern for any science program; and such numbers should be an impetus for change. Through understanding more about the aspects of college STEM education, students and administrators will be better able to manage the challenges of high attrition from STEM programs.

## Differing STEM Faculty Expectations for Success

The lack of consensus among high school and college educators about what introductory STEM courses should entail may also contribute to the difficulties students face in their transition to college. College teachers design their introductory courses with certain assumptions about the academic characteristics of incoming students. High school teachers teach to prepare students for successfully engaging in college STEM courses. According to studies by Daempfle in 2004, Mitchell in 1990, and Razali and Yager in 1994 on college and high school STEM *faculty expectations,* the trend showed that post-secondary instructors had different expectations for incoming student characteristics than those addressed by high school teachers.[30] Differing expectations and preparation at between high school and college may lead to a great divide in the transition, as shown in **Figure 13.2.**

My study examined how well matched high school and college teacher assumptions were about the student characteristics required for success in introductory college biology courses. The results of this study indicated that the two groups of teachers have differing expectations for what is needed for success in college science. College teachers emphasized the importance of mathematics, writing skills, and integrating biology with other subjects, for example. High school faculty instead valued other content areas such as vocabulary knowledge and nomenclature skills (e.g., Latin usage), and dispositions such as self-discipline.[31]

The results imply that high school teachers who believe that the content they teach is important in preparation for college are, in fact, concentrating on areas that college professors do not value highly in college STEM courses. Awareness that this mismatch in preparation may exist should caution the incoming STEM student to beware of their high school preparation and become quickly adept at knowing what their college professors require. Working with upperclass STEM students and gaining a relationship

**Figure 13.2**    Great Wall of China

with faculty are important ways to "learn the ropes" quickly at an institution or in a STEM major.[32]

## Epistemological Assumptions

There is a growing body of evidence that indicates that student scientific *epistemological assumptions,* or how they view science knowing, may affect their academic performance in STEM areas and hence attrition. For instance, the more students believe in the certainty of knowledge, a lower level in the typology of argumentation, the more likely they are to do poorly in STEM majors. These lower-level assumptions interpret tentative science information as absolute, thus leading to a difficulty in comprehending the essence of scientific process—that of disproof of hypotheses and the continual change in accepted science knowledge.[33]

Some studies show that the more college students believe in black and white interpretations of scientific phenomena, the more likely they are to oversimplify information, the poorer their overall academic achievement and confidence in science, and the higher their chances for attrition. Perhaps students with naive epistemological understandings of the nature of science also view science as a chilly climate because they are not "getting it." Students with lower-level, right/wrong beliefs about knowledge are likely to become disenfranchised when confronting the higher-level epistemological demands required by STEM professors, resulting in their attrition.[34]

The majority of STEM dropouts demonstrate a common orientation to science and math ways of knowing: They showed an uncritical acceptance of science and math as factual descriptions of right and wrong answers. Students did not critically consider knowledge and instead possessed only a superficial view of science. Most *nonpersisters* considered science learning the acquisition of a thing. They sought a product that could be obtained and used (a college degree). A utilitarian purpose was the reason for obtaining a science education. This led to eventual student attrition.[35]

We began the text with a philosophical approach to science, one that attempts to bring the reader to higher levels within the typology of argumentation. It is a goal of this text to guide the reader to view science as changing and in flux, continually debated in accordance with the typology of argumentation.

## The Need for Change

In summary, if society expects students to be able to "get excited" about science but the current educational system does not foster such a disposition, there need to be educational changes to improve student achievement and retention. A variety of possible variations may enhance student learning and improve science education.

This chapter clarifies and interprets the interaction of those characteristics of the structure and culture of college science programs that perpetuates high attrition among first-year college STEM majors. The interaction of instructional factors, differing high school and college faculty expectations for entering STEM undergraduates, and epistemological considerations could contribute to a higher dissatisfaction found among STEM majors and the resulting attrition. The need for changes in science education should be addressed in order to enhance science's appeal to U.S. students. This chapter should help students entering college better understand the existing research on college science teaching; it is vital for their success in science programs.

# ■ KEY TERMS

attrition

chilly climate hypothesis

epistemological assumptions

faculty expectations

faculty-student research

non-persister

persister

science, technology, engineering, and
    mathematics (STEM)

underrepresented groups (URG)

# ■ PROBLEMS

1. Describe a good experience you had in a science classroom. What made you "turned on" to science?
2. Describe a bad experience you had in a science classroom. What made you "turned off" to science?
3. What factors do you think lead to retaining and recruiting people into STEM areas? Why?
4. What kinds of students leave science, in your experience? Does this chapter support or refute the research findings?
5. What recommendations would you make, based on the research findings in this chapter, to improve the state of science education?

# ■ REFERENCES

1. Fairweather, J. 2009. Linking Evidence and Promising Practices in Science, Technology, Engineering, and Mathematics (STEM) Undergraduate Education: A Status Report for The National Academies National Research Council Board of Science Education. Center for Higher and Adult Education, Adult and Lifelong Education: Michigan State University. http://www7.nationalacademies.org/bose/Fairweather_CommissionedPaper.pdf. Accessed July 14, 2012.

2. Seymour, E. and Hewitt, N. 1994. *Talking about leaving: Factors contributing to high attrition rates among science, math, and engineering undergraduate majors.* Final report to the Alfred P. Sloan Foundation on an Ethnographic Inquiry at Seven Institutions. Boulder, CO: University of Colorado.

3. Hurtado, Eagan, Chang. 2010. *Degrees of Success: Bachelor's Degree Completion Rates among Initial STEM Majors.* Los Angeles: Education Research Institute; Center for Institutional Data Exchange and Analysis. 2000. *1999–2000 SMET retention report.* Norman, OK: University of Oklahoma.

4. Daempfle, P. 2004. An analysis of the high attrition rates among first-year college science, mathematics, and engineering majors. *Journal of College Student Retention* 5(1):37–52.

5. Astin, A. 1977. *Four critical years.* San Francisco: Jossey-Bass.

6. Astin, A. and Astin, H. 1993. *Undergraduate science education: The impact of different college environments and the educational pipeline in the colleges.* Los Angeles: Higher Educational Research Institute, UCLA.

7. Seymour, E. and Hewitt, N. 1997. *Talking about leaving: Why undergraduates leave the sciences* (p. 35). Boulder, CO: Westview Press.

8. Hilton, R. and Lee, D. 1988. Student interest and persistence in science: Change in the educational pipeline in the last decade. *Journal of College Student Retention* 59 (5):510–526.

9. Daempfle, P. 2000. *Faculty assumptions about the student characteristics required for success in introductory college biology.* Doctoral Dissertation. Accession No. AAI9997550. Albany, NY: The University at Albany.

10. National Research Council. 1996. *National Science Education Standards.* Washington, DC: National Academy of Sciences Press.

11. Champagne, A. and Hornig, L. 1985. *Science teaching.* Washington, DC: American Association for the Advancement of Science.

12. Daempfle, P. 2004. An analysis of the high attrition rates among first-year college science, mathematics, and engineering majors. *Journal of College Student Retention* 5(1):37–52.

13. Seymour, E. and Hewitt, N. 1997. *Talking about leaving: Why undergraduates leave the sciences* (pp. 1–7). Boulder, CO: Westview Press.

14. McShannon, J. and Derlin, R. 2000. *Retaining minority women engineering students: How faculty development and research can foster student success.* Paper presented at the New Mexico Higher Educational Assessment Conference, Las Cruces, NM, February 2000.

15. Seymour, E. and Hewitt, N. 1997. *Talking about leaving: Why undergraduates leave the sciences* (pp. 35–41). Boulder, CO: Westview Press.

16. Daempfle, P. 2003. Faculty assumptions about the student characteristics required for success in introductory college biology. *Bioscene: The Journal of College Biology Teaching,* 28(4):19–33.

17. Seymour, E. and Hewitt, N. 1997. *Talking about leaving: Why undergraduates leave the sciences* (pp. 13–29). Boulder, CO: Westview Press.

18. Ibid, pp. 28–29.

19. Ibid, pp. 40–50.

20. Ibid, pp. 40–45.

21. Ibid, pp. 40–42.

22. Daempfle, P. 2004. An analysis of the high attrition rates among first-year college science, mathematics, and engineering majors. *Journal of College Student Retention* 5(1):37–52.

23. Ibid.

24. Ibid.

25. Ibid.

26. Seymour, E. and Hewitt, N. 1997. *Talking about leaving: Why undergraduates leave the sciences* (pp. 11–29). Boulder, CO: Westview Press.
27. Strenta, C., Elliott, R., Russell, A., Matier, M., and Scott, J. 1994. Choosing and leaving science in highly selective institutions. *Research in Higher Education*, 35:513–537.
28. Daempfle, P. 2004. An analysis of the high attrition rates among first-year college science, mathematics, and engineering majors. *Journal of College Student Retention* 5(1):37–52.
29. McShannon, J., 2001. *Gaining retention and achievement for students program: A faculty development program to increase student success.* ERIC Document Reproduction Service No. ED450627. Available at www.eric.ed.gov.
30. Daempfle, P. 2004. An analysis of the high attrition rates among first-year college science, mathematics, and engineering majors. *Journal of College Student Retention* 5(1):37–52.
31. Ibid.
32. Ibid.
33. Ibid.
34. Ibid.
35. McDade, R. 1988. Knowing the right stuff: Attrition, gender, and scientific literacy. *Anthropology and Education Quarterly,* 19 (2):93–114.

# CHAPTER 14

# Science at Risk

Introduction

The Fall of U.S. Dominance in STEM Areas

The Rise of the Rest

Future of Science Progress in the Twenty-First Century

Improvements to Science Education Are Our Link to Success

Recommendations

## Introduction

Recruiting and retaining qualified people in the sciences are major goals for any advanced society, as described in another chapter. Recent shortages in science-oriented career professionals are the leading cause of a variety of national problems: lack of qualified medical doctors, particularly in primary care areas; deficiencies in or underprepared science teachers; a major nursing shortage, predicted to worsen in coming years; lack of science researchers in medically-oriented fields such as the pharmaceutical industry; and few physical scientists for military technology development.[1]

As a remedy, the U.S. has outsourced to meet these needs by importing science talent from abroad. In fact, less than half of all Ph.D. candidates in STEM in U.S. universities are American students. Foreign born nationals often use the American post-secondary school system to get an education and then return abroad. In fact, over 60% of engineering degrees awarded last year were to non-citizens. The U.S. Patent Office reported that immigrants invent patents at roughly double the rate of non-immigrants. This is why a 1% increase in immigrants with college degrees leads to a 15% rise in the production of patents. Immigrants are responsible for a good deal of job growth, with 52% of silicon valley tech companies started by foreign born STEM majors.[2] While science is universal, this is clearly a threat to our national security. When our nation cannot sustain a populace of scientifically productive individuals to suit the nation's needs, concern is warranted. In the U.S., we certainly are not doing our part to educate our citizenry to contribute to international science. Science in our world is thus at risk.

## The Fall of U.S. Dominance in STEM Areas

The U.S. has operated under the assumption that it will remain a superpower forever, given its vast abundance of resources. The country was able to win wars and outperform industrially due to technological advances. Improvements in technology have helped every civilization to win nearly every war and every battle throughout history. For instance, the Mongols of central Asia gained a vast empire through using technology along with clever military tactics (1206–1324). *Siege engines* were developed at the site of battle by conquered engineers who the Mongols imprisoned. These machines destroyed the walls of a fortified city, enabling *kharask* tactics, which used prisoners as shields to march into the enemy fort. Thus, take over of previously impregnable fortresses were based on both technology and harsh military measures. Genghis Khan was thus able to expand the Mongol Empire, conquering nearly all of continental Asia, the Middle East and parts of eastern Europe. The development of siege tactics using gunpowder led to modern day cannon artillery. In more modern times, nuclear fission ended WWII rather quickly and smart rocketry rapidly defeated U.S. enemies in the Persian Gulf Wars.

The question always at hand is, "What is the next technology to supplant our existing advantages?" One can speculate about possibilities from STAR WARS missile defenses to better technological controls of bombs. Further, something less feasible but strategically coveted, like invisibility, comes to mind. Imagine if one were able to become invisible. Then simple infiltration of our defenses could lead to destruction of the nation, as intimated in **Figure 14.1**. While seemingly far-fetched, its point should be taken that technology and sciences are vital to our security and economy.

The trends in our national science and mathematics comparatives are a most disconcerting set of data. The *Organization for Economic Co-operation and Development (OECD)* reports that the country's 15-year-olds rank 17th in the world in science and 25th in mathematics. Our college graduation rates rank 12th and our infrastructure ranks 23rd in the world. Only a few decades ago, when STEM areas were more respected by our society, the U.S. ranked in the top notch positions. We are currently 27th in life expectancy and 18th in diabetes rates but #1 in obesity rates and debt.[3] As a nation we should strive for better rankings in the science, mathematics, and engineering areas. Our people are better than the numbers show. What a shame.

The *No Child Left Behind Act (NCLB)* of 2001 was an attempt by the U.S. government's Bush administration to ameliorate the deteriorating standards in academics. In theory, it was an appropriate measure to rein in a lack of centralized

© nokhoog_buchachon/ShutterStock, Inc.

**Figure 14.1**    The Invisible Spy

control over education and institute standards across the U.S. In this way, the work accomplished was an excellent first step in establishing standards for success and making teachers more accountable for learning. Standards were enacted and tested and great care was taken to ensure the tests fit what students were supposed to learn. It was a bold move to bring the nation's STEM areas back to dominance.

Unfortunately, the act fell short of its intended goals. Measured and sustained progress could have been implemented had the Act had further support to include more science areas. It lacked popular political support and serious financial extension. However, there are some real successes. Science students in Massachusetts improved in their scores to match most of Europe and other states improved significantly on many academic measures.

My involvement in NCLB as a science consultant and advisor since its inception was at many levels, from standard realignment to reviewing and creating assessments appropriate for success in college science. I am currently the expert reviewer for state-level annual secondary biology assessment. It is with this experience that I can speak about the importance of maintaining standards but also having the administrative support to align instruction to do this.

However, there were major problems with the Act. Students were made to feel greater test anxiety and institutions focused primarily on teaching to the test. Institutions were held back by low grades and school district funding was on the line, making testing a stressful endeavor. Instead, the Act should have had more financial and structural support for addressing individual students' needs to help them achieve higher scores.

As mentioned in other chapters, changes in educational structure and culture such as improvement of teacher qualifications and academic tutoring services, should have been more seriously implemented in conjunction with the standards-based reforms of the NCLB Act. The Act was reduced to, in large measure, a game of schools competing for funding. Where is science education headed next in the absence of support for NCLB?

Currently, the trend in standards-based school assessment is to have every person reach a level of science achievement under the No Child Left Behind Act of 2001 and *Race to the Top* of 2009. Both administrative plans purport that all students can universally learn science to a level that will help them make life decisions better and function in modern employment. Learning science and the scientific process is a noble and humane goal for improving peoples' lives, a focus of this text.

However, there are currently no clear national initiatives for improving STEM education, even in the face of the many deficiencies in science related professions. In fact, the current U.S. administration and its Congress are set to cut funding to areas such as education, scientific research, air-traffic control, NASA, infrastructure, and alternative energy research.[4] These cuts will help provide immediate budget relief, but at high costs to society at large.

Other nations including China, Germany, and South Korea are making large investments into science, mathematics, and engineering areas. The U.S is cutting these areas while simultaneously increasing subsidies for consumption of foreign-made goods. Demand side economics is being practiced, whereby monies are being given to the public to spend more on consumer products and small ticket items that fail to create American jobs because these products are made at lower cost abroad.

This is the opposite of what should be done to drive economic growth. New technologies and industries, as developed by STEM professionals, have always been the main creator of jobs and economic growth. The TV, radio, and car were developed in this way and changed the economy. Today's economy was built up by policies developed almost 50 years ago, under the Eisenhower administration. To illustrate, engineering projects and research built the interstate highway system in the 1950s and early 1960s. It only takes a drive through the U.S. to realize that our infrastructure is failing to meet the nation's new demands. In the past decade, drivers on roads increased over 10% while road building across the U.S. increased only 1%. In this same time period, time spent driving increased 35% and the number of cars and trucks on the roads increased 17%.[5] Few large-scale additions of rails or highways were implemented in the last half-century and the nation is suffering. Tractor trailers dominate the interstate roads, and the rail system is inefficient, failing to meet the transportation needs of the nation. Competition continues to drive scientific research in modern society. China has recently developed a very high speed rail system, recently testing a train at 310 mph. It would be fantastic for U.S. infrastructure, rivaling air transit which averages 450 mph, and outperforming our existing trains which average a slow 65 mph (with multiple malfunctions). These are only small examples of how neglected STEM sections of the economy are leading to national weaknesses.

## The Rise of the Rest

Technology has enabled the U.S. to produce more goods and services with fewer people and more effectively. For example, one tractor is more efficient than 10 men in a field. That said, other nations are catching up by playing the same game as us. Harvard historian Niall Ferguson, in his book, *Civilization: The West and the Rest,* places this in historical context:

> For 500 years the West patented six killer applications that set it apart. The first to download them was Japan. Over the last century, one Asian country after another has downloaded these killer apps—competition, modern science, the rule of law and private property rights, modern medicine, the consumer society, and the work ethic. Those six things are the secret sauce of Western civilization.[6]

While our nation's dominance holds science, technology, and medicine as its main drivers, it is unclear that we will remain on top given the state of science education today.

## Future of Science Progress in the Twenty-First Century

Increases in attrition rates among STEM majors have produced a variety of deleterious effects for the society. The future success of our society lies in its ability to compete internationally in STEM areas. Improving the education of our students in STEM areas is a key to the driving force behind our future success. This section and the next summarize and make recommendations about the interaction of those characteristics of the structure and culture of STEM programs that perpetuate high loss rates. As stated in another chapter, losses take place particularly during the first year of college science. This is the time when it is most important to get qualified students to enter and remain in a STEM major. The interaction of instructional factors, differing high school and college faculty expectations for entering STEM undergraduates, and many aspects of both science teaching were found to contribute to a higher dissatisfaction among STEM majors as compared with non-STEM majors and the resulting attrition. Improvements in STEM education were shown to be critical for improving science literacy and future employment in STEM careers.[7]

## Improvements to Science Education Are Our Link to Success

As seen in the previous chapter, improvements in science education are likely to lead to better recruitment and retention of capable people in science. How can this be accomplished? To reduce student loss rates in college STEM courses, the research clearly points to a need to change the structure of the post-secondary STEM classroom. To accomplish this, undergraduate STEM instruction should shift from simple knowledge transmission to actively and cooperatively engaging students. As seen in previous chapters, active student involvement in critical thinking and collaboration results in higher student achievement along with important outcomes in science reasoning and thinking. Engaging students in lectures, structuring assessment practices to include cooperative learning strategies, and increasing faculty involvement would improve student attitudes, achievement, and retention.

The mere presentation of content in introductory college science courses without giving students opportunities for seeing how that knowledge fits within the whole curriculum does not further the kind of collateral, integrative learning advocated by national and state secondary standards in STEM areas. A program of retraining educators to understand the characteristics of their incoming students in college science is very important in creating the kinds of changes advocated in previous chapters. Willingness to accept changing teaching methods requires a change in the culture of college science.

Although scientific factual knowledge is needed in the vertically-structured sciences, it is only insofar as it can be used in higher order cognitive processes, which compare and connect the phenomena of a variety of academic areas.[8] *John Dewey* (1858–1952), an early educational researcher, cited the importance of a principle of continuity between different areas of knowledge. This is reflected by my recommendations for content in post-secondary STEM courses, whereby students engage in collateral, synthetic learning. This fosters what

Dewey considers "the most important attitude that can be formed[,] . . . that of a desire to go on learning."[9]

To reach college science expectations, content learning demands for college STEM students should thus "be considered intrinsically multidisciplinary. Student learning is enhanced when we are able to help students see the relationships among the sciences, and between science and mathematics, the humanities, social sciences, and the arts."[10] Rather than focusing on what amount of content is necessary, the organization of that content around themes, issues, or projects can enrich the student learning. A student's view that the sciences are not separable from other areas of knowing will be better developed. Students will then be able to reason about science in a more holistic way.

Despite its limitations, traditional lectures are the most common form of instruction in introductory STEM courses.[11] I would recommend a change from the lecture-oriented instruction. However, it may be practically unavoidable given the institutional policies on limiting instructors and maintaining high introductory course enrollments. A commitment to STEM courses, financially and philosophically, by administrations, is essential to lowering class sizes and appreciating the need for reform in STEM education.

A look to the literature on teaching and learning that contains instructional strategies to enhance student learning in lecture settings is also beneficial. For example, some advice for science instruction may include: Use paradoxes and apparent contradictions to engage students; make connections with other courses and everyday phenomena; begin each class with something familiar to students; delivery affects student motivation (e.g., eye contact, enthusiasm); ask *divergent questions* (having more than one answer) over *convergent questions* having only one answer.[12]

When considering the final point, the type of questions asked by science teachers are important to the kinds of reasoning processes students are encouraged to use and the kind of excitement brought to the lecture. For example, a divergent type question such as "*Why* do birds produce uric acid as an excretory product?" would elicit a much more exciting and reasoned answer than the convergent, "*What* do birds produce as an excretory product?" Consider the first divergent question. A student must elaborate on their prior knowledge that uric acid is a precipitate from chemistry class, that urea produced by humans would kill the bird embryo because it is soluble in the water within the shell (concepts from excretion in introductory biology), and that evolutionarily this is beneficial (a sociohistorical principle) because it will not kill the embryo like urine would. Alternate working hypotheses may be developed through class discussion focusing on, for example, the conservation of water in excreting uric acid instead of urea. The reasoning moves beyond the mere acquisition of the fact that birds make uric acid and enhances the kind of collateral thinking that would create a more interesting and less "chilly" classroom, with the student creating knowledge collaboratively.[13]

## Recommendations

An active national research program, perhaps tracking nonpersisting students in science and exploring their reasons for leaving, would be beneficial to improving retention. National standardization of testing and content standards might be an important step in improving scientific literacy. Action at the institutional level would follow. Institutional action is supported by this text in accordance with the National Action Council for Minorities in Engineering's 1996 conceptual framework for treating student retention as a top issue.[14]

Because a mismatch in student preparation between secondary and post-secondary STEM programs is established, safeguards against student loss should be implemented at national, state, and institutional curricular and instructional levels. These might include: remediative courses or workshops to lessen possible academic deficiencies caused by preparation discrepancies, counseling services to help students cope with academic adjustments during their transition to college STEM courses, recruitment of science faculty willing to participate in attrition research, and fostering communication among institutions to improve the congruence of student preparation at transition stages in their education.

The greatest challenge is to excite, recruit, and retain qualified students into the sciences. That is the goal of this text and it is hoped that the content knowledge, philosophical and historical base, along with the mathematics and critical thinking skills developed through this text, has developed the readers' science literacy and an excitement about science. In this way, readers will have the desire to go on learning science and using their science understanding to better their lives.

## ■ KEY TERMS

| | |
|---|---|
| convergent question | Organization for Economic |
| Dewey, John | Co-operation and Development |
| divergent question | (OECD) |
| kharask | Siege Engine |
| No Child Left Behind Act (NCLB) | Race to the Top |

## ■ PROBLEMS

1. What do you think are the causes of the decline in U.S. rankings in STEM test scores?
2. Are there other instructional or institutional recommendations you would make to enhance retention and recruitment to STEM areas?

3. Develop a divergent question regarding *photosynthesis*. Then construct an equivalent convergent type question. List three differences between the two in what you would hope to achieve if you were the science teacher asking the questions.

4. What, in your opinion, is the most important socioeconomic link to science education in the U.S.? In world competitiveness?

5. Imagine a new technology that could be developed to threaten U.S. security. Do you think that your imagination is a possibility?

6. Which of the changes in U.S. world rankings listed in this chapter bother you the most? Explain.

## ■ REFERENCES

1. Daempfle, P. 2004. An analysis of the high attrition rates among first-year college science, mathematics, and engineering majors. *Journal of College Student Retention* 5(1):37–52.

2. Lehrer, J. 2011 (May 14). Editorial Why America needs immigrants. *The Wall Street Journal*, http://online.wsj.com/article/SB10001424052748703730804576313490871429216.html. retrieved July 10, 2012.

3. Zakaria, F. Are America's best days behind us? *Time Magazine* Mar. 3, 2011.

4. Ibid.

5. Ricardo, Martinez. 1997. Statement of the Honorable Ricardo Martinez, M.D. administrator, *National Highway Traffic Safety Commission*, U.S. House of Representatives, Subcommittee on Surface Transportation, Washington, D.C.

6. Zakaria, F. Are America's best days behind us? *Time Magazine* Mar. 3, 2011.

7. Daempfle, P. 2004. An analysis of the high attrition rates among first-year college science, mathematics, and engineering majors. *Journal of College Student Retention* 5(1):37–52.

8. Ibid.

9. Dewey, J. 1938. *Experience and education* (p. 48). New York: Collier Macmillan Publishers.

10. National Academy Press. 1997. *Science teaching reconsidered: A handbook* (pp. 10–11). Washington, DC: Committee on Undergraduate Science Education.

11. Ibid, p. 7.

12. Ibid, pp. 2–9.

13. Daempfle, P. 2004. An analysis of the high attrition rates among first-year college science, mathematics, and engineering majors. *Journal of College Student Retention* 5(1):37–52.

14. National Academy Press. 1997. *Science teaching reconsidered: A handbook*. Washington, DC: Committee on Undergraduate Science Education.

# Glossary

**acid**: A chemical compound with a pH less than 7. Donates hydrogen ions to a solution.

**acupuncture**: An alternative medicine methodology that involves the practice of inserting needles into the body to reduce pain or induce anesthesia.

**aerobic**: Living or occurring only in the presence of oxygen.

**affective reasoning strategies**: Refers to critical thinking method that involves feelings, emotions, and attitudes. Compare with cognitive reasoning strategies.

**alchemy**: A science and philosophy from the Middle Ages that attempted to transform metals into gold; early chemistry.

**altruistic**: Unselfish concern for the welfare of others; selflessness.

**anabolism**: The synthesis in living organisms of more complex substances from simpler ones. Compare with catabolism.

**Analysis of variance (ANOVA) test**: A test that identifies and isolates the sources of variation during hypothesis testing.

**aneurism**: A localized, blood-filled dilation of a blood vessel caused by disease or weakening of the vessel's wall.

**angioplasty**: minimally invasive procedure that open ups clogged arteries.

**Aquinas, Thomas**: Medieval theologian (1225–1274) who integrated the ancient views of Aristotle into a harmony with Christian ideology, emphasizing that the Earth was the center of the universe and that humans were special creations in that center.

**argument**: A discussion in which disagreement is expressed; a debate.

**Aristotle**: The great Greek philosopher of the fourth century (384–322 B.C.) who formed the major tenets of natural philosophy.

**arteriosclerosis (coronary artery disease)**: A chronic disease in which thickening, hardening, and loss of elasticity of the arterial walls result in impaired blood circulation.

**asthenosphere**: A zone of the earth's mantle that lies beneath the lithosphere and consists of several hundred kilometers of deformable rock.

**astrology**: The study of the positions and aspects of celestial bodies in the belief that they have an influence on the course of natural earthly occurrences and human affairs; early astronomy.

**atom**: The smallest component of an element, having all the characteristics of that element and consisting of a dense, central positively charged nucleus surrounded by a system of electrons.

**atomic physics**: Scientific study of the structure of the atom, its energy states, and its interaction with other particles and fields.

**attrition**: A reduction or decrease in numbers, size, or strength of students in an educational program.

**base:** Ionic compound that breaks apart to form a negatively charged hydroxide ion ($OH^-$) in water. Absorbs hydrogen ions from a solution.

**behavioral genetics:** The study of interactions between genetic factors and environment in determining behavior.

**bias:** Inclination to present or hold a partial perspective at the expense of (possibly equally valid) alternatives.

**Big Bang theory:** A theory of cosmology holding that the expansion of the universe began with a gigantic explosion between 12 and 20 billion years ago.

**biogeochemical cycling:** The cycling of chemical elements required by life between the living and nonliving parts of the environment.

**biology:** The science of life and of living organisms, including their structure, function, growth, origin, evolution, and distribution.

**blinded study:** A study done in such a way that the patients or subjects do not know what treatment they are receiving to ensure that the results are not affected by a placebo effect (the power of suggestion).

**Boyle, Robert:** Irish natural philosopher (1627–1691) who made important contributions to chemistry and physics and established himself as a leading figure in the scientific revolution.

**Boyle's law:** Describes the inversely proportional relationship between the absolute pressure and volume of a gas, if the temperature is kept constant within a closed system.

**Bruno, Giordano:** An Italian monk, philosopher, mathematician, and astronomer (1548–1600) who extended Copernicus's ideas and publicly taught that the universe was infinite and had no surfaces and no end. He was burned at the stake for heresy.

**calculus:** A branch of mathematics that looks at things that change over time. It tries to say what type of change it is and how big the change is using functions at the exact moment the change is taking place.

**calendar year:** Explanation of the history of the universe condensed into one year; first proposed by Carl Sagan.

**Cannon, Walter Bradford:** Harvard physiologist and World War I medical doctor (1871–1945) who pioneered the idea of homeostasis.

**capitalism:** An economic system that is based on private ownership of the means of production and the creation of goods or services for profit.

**carbohydrate:** Any member of a large class of chemical compounds that includes sugars, starches, cellulose, and related compounds.

**Cartesian dualism:** The view that mind and body are two separate substances; the self is as it happens associated with a particular body, but is self-subsistent, and capable of independent existence.

**catabolism:** The breaking down in living organisms of more complex substances into simpler ones, with the release of energy. Compare with anabolism.

**cell:** The basic structural and functional unit of any living thing. Each cell is a small container of chemicals and water wrapped in a membrane.

**central tendency:** The degree of clustering of the values of a statistical distribution that is usually measured by the arithmetic mean, mode, or median.

**chemistry:** The branch of physical science concerned with the composition, properties, and reactions of substances.

**chilly climate hypothesis:** The idea that an unapproachable or "cold" attitude of faculty results in attrition in the classroom.

**chiropractics:** Branch of alternative medicine grounded in the principle that the body can heal itself when the skeletal system is correctly aligned and the nervous system is functioning properly. From Greek words meaning done by hand.

**cochlea:** A spiral tube forming part of the inner ear, which is the essential organ of hearing.

**cognitive reasoning strategies:** Refers to critical thinking method that involves intellectual skills, such as interrelating ideas, memorizing, and developing hypotheses. Compare with affective reasoning strategies.

**cold fusion:** A hypothetical form of nuclear fusion occurring without the use of extreme temperature or pressure.

**collaborative instruction:** Nontraditional teaching approach accomplished in a joint intellectual effort between faculty and students.

**communism:** A theoretical economic system characterized by the collective ownership of property and by the organization of labor for the common advantage of all members.

**comparative anatomy:** The study of similarities and differences in the anatomy of different organisms.

**concrete reasoning:** Reasoning characterized by a predominance of actual objects and events and the absence of concepts and generalizations. Compare with critical reasoning.

**continental drift:** Theory which states that parts of the Earth's crust slowly drift atop a liquid core, resulting in a constant state of movement.

**control variable:** A variable that is held constant or whose impact is removed in order to analyze the relationship between other variables without interference.

**convergent junction:** A tectonic boundary where plates crash or crunch together. Compare with divergent junction.

**convergent question:** Question that represents the analysis and integration of given or remembered information, leading to an expected end result or answer.

**Copernicus, Nicolaus:** A Renaissance astronomer and mathematician (1473–1543) who was the first person to formulate a comprehensive heliocentric cosmology that proposed that the Sun was stationary in the center of the universe and the Earth revolved around it.

**Copernican revolution:** Refers to the paradigm shift away from the Ptolemaic model of the heavens, which postulated the Earth at the center of the galaxy, toward the heliocentric model with the Sun at the center of our solar system. It was one of the starting points of the Scientific Revolution of the sixteenth century.

**core:** The central part of the Earth beneath the mantle, consisting mainly of iron and nickel, which has an inner solid part surrounded by an outer liquid part.

**correlation:** Mutual relationship or connection; the degree of relative correspondence, as between two sets of data.

**correlation coefficient:** A statistical measure of the interdependence of two or more random variables.

**counterargument:** An argument, or a reason or line of reasoning, given in opposition to another argument.

**Crick, Francis:** English scientist (1916–2004) who is credited with the discovery of the structure of the DNA molecule, along with James Watson, at Cambridge University in 1953, and was jointly awarded the 1962 Nobel Prize for Physiology or Medicine.

**critical reasoning:** Reasoning that is clear, rational, open-minded, and informed by evidence. Compare with concrete reasoning.

**critical review of research:** Evaluation of a topic from every angle for inconsistencies in reported research results, flaws in designs, identification of alternative routes to study topics, and critique of mathematical findings.

**critical thinking:** The ability to judge a claim and change one's reasoning about it, if deemed necessary.

**dark matter:** A type of matter hypothesized to account for a large part of the total mass in the universe. It cannot be seen directly with telescopes; instead, its existence and properties are inferred from its gravitational effects on visible matter, radiation, and the large-scale structure of the universe.

**Darwin, Charles:** English naturalist (1809–1882) who is credited with the scientific theory of evolution, which resulted from a process that he called natural selection.

**deduction:** The process of reasoning from one or more general statements regarding what is known to reach a logically certain conclusion. Contrast with induction.

**deductive methodology:** A tenet of modern science philosophy that states that conclusions need to logically flow from any premise.

**dependent variable:** A characteristic, number, or quantity that is measured in an experiment and what is affected during the experiment.

*De Revolutionibus:* Seminal introduction of the heliocentric theory by the Renaissance astronomer Nicolaus Copernicus (1473–1543).

**Dewey, John:** An American philosopher, psychologist and educational reformer (1859–1952) whose ideas have been influential in education and social reform.

**dialectic thinking:** Discussion and reasoning by dialogue as a method of intellectual investigation; specifically, the Socratic techniques of exposing false beliefs and eliciting truth.

**dialogic argumentation:** Considers science a dialogue in which discussion leads to an exposing of contrasting ideas to facilitate reflective thinking.

**dichotomy:** A division into two especially mutually exclusive or contradictory groups or entities.

**diet:** Food and drink habitually consumed by a person or animal.

**diffusion:** The spontaneous migration of substances from regions where their concentration is high to regions where their concentration is low.

**diploid:** Double or twofold. Having a pair of each type of chromosome, so that the basic chromosome number is doubled. Contrast with haploid.

**discussion:** Consideration of a question in open and usually informal debate.

**disinterest:** Freedom from selfish bias or self-interest; impartiality.

**dissertation:** A lengthy, formal treatise, especially one written by a candidate for the doctoral degree at a university; a thesis.

**distribution:** The set of possible values of a random variable, or points in a sample space, considered in terms of new theoretical or observed frequency.

**diuretic:** Substance or drug that tends to increase the discharge of urine. Diuretics are used in the treatment of high blood pressure, edema, and other medical conditions.

**divergent junction:** A tectonic boundary where plates spread apart. Compare with convergent junction.

**divergent question:** A question with no specific answer, but rather exercises one's ability to think broadly about a certain topic.

**double-blinded study:** A medical study in which both the subjects participating and the researchers are unaware of when the experimental medication or procedure has been given.

**dowsing:** A method of divination used to locate water, metals, gemstones, and hidden objects with the aid of simple handheld tools or instruments.

**egocentricity:** Having or regarding the self or the individual as the center of all things rather than society. Contrast with sociocentricity.

**Einstein, Albert:** German theoretical physicist (1879–1955) who developed the theory of general relativity, effecting a revolution in physics. For this achievement,

Einstein is often regarded as the father of modern physics and he was awarded the 1921 Nobel Prize in Physics.

**electron:** A negatively charged subatomic particle. It can be either free (not attached to any atom), or bound to the nucleus of an atom.

**element:** A substance consisting of atoms which all have the same number of protons.

**empirical method:** The approach of using a collection of data to base a theory or derive a conclusion.

**empiricism:** The search for knowledge by observation and experiment; a disregarding of scientific methods and relying solely on sensory experience.

**Enlightenment:** A philosophical movement of the eighteenth century that emphasized the use of reason to scrutinize previously accepted doctrines and traditions and that brought about many humanitarian reforms.

**enrichment:** The act of making fuller or more meaningful or rewarding; the act of obtaining increased amounts of radioactive material that may be used in nuclear reactors or weapons.

**epistemological assumption:** An understanding that all knowledge is based of assumptions that cannot be proven.

**epistemology:** The branch of philosophy that studies the nature of knowledge, its presuppositions and foundations, and its extent and validity.

**estimated omega squared:** An estimate of the dependent variance accounted for by the independent variable in the population for a fixed effects model.

**ethics:** A branch of philosophy that involves systematizing, defending, and recommending concepts of right and wrong behavior.

**eukaryote:** A single-celled or multicellular organism whose cells contain a distinct membrane-bound nucleus. Contrast with prokaryote.

**eusocial system:** A cooperative group in which usually one female and several males are reproductively active and the nonbreeding individuals care for the young or protect and provide for the group.

**evolution:** A theory that the various types of animals and plants have their origin in other preexisting types and that the distinguishable differences are due to modifications in successive generations.

**experiment:** An operation or procedure carried out under controlled conditions in order to discover an unknown effect or law, to test or establish a hypothesis, or to illustrate a known law.

**experimenter bias:** A process where the scientist(s) performing the research influence the results, in order to portray a certain outcome.

**extraneous variable:** An undesirable variable that may bear any effect on the behavior of the subject being studied.

**extrapolation:** Technique of inferring something unknown from the known. Compare with interpolation.

**F-ratio:**  A value used in determining whether the difference between two variables is statistically significant or stable.

**faculty expectations:**  Refers to the anticipated level of preparedness that instructors have for incoming students.

**faculty-student research:**  Academic relationship between instructor and student in the research setting; it is an area where positive mentoring can be established.

**falsification:**  The act of producing something that lacks authenticity and passing it off to other people as authentic.

**formal reasoning:**  The use of reason, especially to form conclusions, inferences, or judgments or the evidence or arguments used in thinking or argumentation.

**Franklin, Rosalind:**  A British biophysicist and X-ray crystallographer (1920–1958) who is best known for her work on the X-ray diffraction images of DNA that led to discovery of DNA double helix.

**fraud:**  A deception deliberately practiced in order to secure unfair or unlawful gain.

**friction:**  A force on objects or substances in contact with each other that resists motion of the objects or substances relative to each other.

**fusion:**  The act or process of fusing or melting together; a union.

**Galilei, Galileo:**  An Italian physicist, mathematician, astronomer, and philosopher (1564–1642), who played a major role in the Scientific Revolution when he was put on trial for popularizing the view that the sun, and not the Earth, was the center of planetary motion.

**geocentric model:**  A paradigm that places the Earth at the center of the universe. Common in ancient Greece after the discovery of the approximately spherical shape of Earth, it was believed by both Aristotle and Ptolemy. Contrast with heliocentric model.

**general theory of relativity:**  The geometric theory of gravitation that proposes that matter causes space to curve. This theory was published by Albert Einstein in 1916 and is the current description of gravitation in modern physics.

**genetic epistemology:**  Theory by Jean Piaget that our understanding of knowledge develops in actual people (children) instead of simply as an abstract philosophy based on our adult intuitions about knowledge.

**geology:**  The branch of science comprising the study of solid Earth, the rocks of which it is composed, and the processes by which it evolves.

**German university system:**  In the first decade of the nineteenth century, German universities were the first to offer natural science as a field of study.

**germ theory:**  Theory invented by Louis Pasteur that states that specific microscopic organisms are the cause of specific diseases.

**Giant Impact theory:**  Hypothesis states that the Moon was formed out of the debris left over from a collision between the Earth and a Mars-sized body, sometime around 4.5 billion years ago.

**glucagon:** A hormone produced by the pancreas that stimulates an increase in blood sugar levels, thus opposing the action of insulin.

**glucose:** A simple sugar that is the principal circulating sugar in the blood and the chief source of energy of the body.

**Gold Foil experiment:** An experiment performed by Ernest Rutherford and his associates in 1911 to determine the layout of the atom. Showed that atoms are composed of mostly empty space.

**group selection:** Describes natural selection operating between groups of organisms, rather than between individuals. Compare with individual selection.

**groupthink:** The practice of approaching problems or issues by consensus of a group rather than by individuals acting independently. In this situation, popularity determines knowledge and truth.

**glycogen:** The principal storage form of glucose in animal cells. In humans and other vertebrates, it is found primarily in the liver and muscle tissue.

**haploid:** Having the same number of sets of chromosomes as a germ cell (sperm or egg) or half as many as a somatic cell (body cell) or having a single set of chromosomes. Contrast with diploid.

**haplodiploidy:** A sex determination system in which males develop from unfertilized eggs and are haploid, and females develop from fertilized eggs and are diploid.

**HeLa cells:** A cell type in an immortal cell line used in scientific research. It is the oldest and most commonly used human cell line, first isolated in 1951 from the cervical cancer cells of a young African-American woman, Henrietta Lacks.

*Helicobacter pylori:* A bacterium that causes stomach inflammation and ulcers in the stomach and duodenum.

**heliocentric model:** The astronomical model in which Earth and planets revolve around a stationary Sun at the center of the universe. Contrast with geocentric model.

**Hermetic doctrine:** States that all matter contains a divine spirit and that divine spirit needs to be studied in order to be unleashed to give knowledge. Based on the writings of Hermes Trismegistus.

**historical geology:** A subdiscipline of human geography concerned with the geographies of the past and with the influence of the past in shaping the geographies of the present and the future.

**homeostasis:** The tendency of a system, especially the physiological system of higher animals, to maintain internal stability, owing to the coordinated response of its parts to any situation or stimulus that would tend to disturb its normal condition or function.

**honesty:** A facet of moral character and denotes positive, virtuous attributes such as integrity, truthfulness, and straightforwardness along with the absence of lying, cheating, or theft.

**hormone:** A chemical substance produced in the body that controls and regulates the activity of certain cells or organs.

**Huxley, Thomas Henry:** A self-educated English biologist (1825–1895) who was one of the first supporters of Darwin's theory of evolution and who did original research in zoology and paleontology on his own in support of natural selection.

*Hymenoptera:* One of the largest orders of insects, comprising the ants, bees, wasps, and sawflies, among others.

**hypothesis:** A tentative explanation for an observation, phenomenon, or scientific problem that can be tested by further investigation.

**igneous rock:** Rock produced under conditions involving intense heat, as rocks of volcanic origin or rocks crystallized from molten magma.

**ill-structured question:** Question that has the possibility of more than one acceptable answer.

**independent variable:** A variable that stands alone and isn't changed by the other variables you are trying to measure.

**individual selection:** A natural or artificial process that favors or induces survival and perpetuation of one kind of organism over others that die or fail to produce offspring. Compare with group selection.

**induction:** A method of reasoning that moves from specific instances to a general conclusion. Contrast with deduction.

**inertia:** A property of matter by which it remains at rest or in uniform motion in the same straight line unless acted upon by some external force.

**informal reasoning:** A broad term for any of the various methods of analyzing and evaluating arguments used in everyday life.

**inorganic chemistry:** The study of the chemistry of materials from nonbiological origins.

**inquiry-based instruction:** A student-centered and teacher-guided instructional approach that engages students in investigating real-world questions that they choose within a broad thematic framework.

**insulin:** A natural hormone made by the pancreas that controls the level of the sugar glucose in the blood.

**interpolation:** Estimation of an unknown quantity between two known quantities (historical data), or drawing conclusions about missing information from the available information. Compare with extrapolation.

**intraspecific:** Occurring within a species or involving members of one species.

**invisible community:** A group of people with common interests who are not easily seen; they are generally hidden.

**Kepler, Johannes:** A German Hermetic scholar, mathematician, astronomer, and astrologer (1571–1630). A key figure in the 17th century scientific revolution, he is best known for his eponymous laws of planetary motion, which led to a view that the universe could be explained mathematically and that there is an order in the way the physical world operates.

**kinetic energy:** Energy of motion.

**King, Patricia and Kitchener, Karen:** Proposed the Reflective Judgment Model in the 1970s, a model that better explains how students develop arguments and judgments.

**laissez-faire economics:** Doctrine that an economic system functions best when there is no interference by government.

**large diameter nerve fibers:** Peripheral nerve fibers with a high conduction velocity.

**Little Ice Age:** A period of cooling that was first recorded around 1300 and extended through to the mid 1800s, which brought about great environmental change.

**lipid:** Any of a group of organic compounds consisting of the fats and other substances of similar properties.

**lithosphere:** The outer part of the Earth, consisting of the crust and upper mantle, approximately 100 km (62 mi) thick.

**logic:** The study of the principles of reasoning, especially of the structure of propositions as distinguished from their content and of method and validity in deductive reasoning.

**macro-level biology:** The study of living organisms at or on a level that is large in scale or scope.

**macromolecule:** A very large molecule, such as a polymer or protein, consisting of many smaller structural units linked together.

**mantle:** The layer in the structure of the Earth that lies directly under the crust. The term is also applied to the structure of other planets.

**matter:** Any substance which has mass and occupies space.

**Maxwell, James:** A Scottish physicist and mathematician (1831–1879) whose most prominent achievement was analyzing the relationship between light, magnetism, and electricity to lay the foundations for the technological and electronics boom of the twentieth century.

**mean:** A numerical value that in some sense represents the central value of a set of numbers; an average. Statistics is based on use of the mean.

**mechanical physics:** Sub-field of physics that is concerned with the set of physical laws describing the motion of bodies under the action of a system of forces.

**median:** One type of central tendency measure, found by arranging the values in order and then selecting the one in the middle.

**metacognition:** "Thinking about thinking," reflecting on one's personal thought processes.

**metaknowledge:** "Knowledge about knowledge," conceptually any definition of knowledge applied to mastering, mapping, or "knowing" a region of knowledge, with scope up to and including all knowledge.

**metamorphic rock:** Rock that arises from the transformation of existing rock types.

**metaphysics:** The branch of philosophy that examines the nature of reality, including the relationship between mind and matter, substance and attribute, fact and value.

**metastasis:** Transmission of disease from one organ or part of the body to another not directly connected with it.

**microbiology:** The branch of biology that deals with microorganisms and their effects on other living organisms.

**micro-level biology:** The study of living organisms at a very small, or microscopic level.

**mode:** One type of central tendency measure; calculated as the number that appears most often in a set of numbers.

**mole:** The molecular weight of a substance expressed in grams.

**monozygotic twins:** A type of twins derived from a single (mono) egg (zygote).

**multiplicity:** The number of times a member of a multiset appears in the multiset.

**nanotechnology:** The study of manipulating matter on an atomic and molecular scale.

**natural history:** The study and description of organisms and natural objects, especially their origins, evolution, and interrelationships.

**naturalistic observation:** A research tool in which a subject is observed in its natural habitat without any manipulation by the observer.

**natural philosophy:** A term applied to the objective study of nature and the physical universe before the development of modern science.

**natural selection:** The process in nature by which, according to Darwin's theory of evolution, only the organisms best adapted to their environment tend to survive.

**natural world:** Refers to the phenomena of the physical world, and also to life in general; synonymous with "nature."

**nature vs. nurture:** Debate that concerns the relative importance of an individual's innate qualities ("nature") versus personal experiences ("nurture") in determining or causing individual differences in physical and behavioral traits.

**negative feedback:** A type of feedback during which a system responds so as to reverse the direction of change.

**neoplatonism:** A school of philosophy developed by Plotinus in Rome, based on modified Platonism, and postulating a single source from which all forms of existence emanate and with which the soul seeks mystical union.

**neutron:** A subatomic particle contained in the atomic nucleus.

**Newton, Isaac:** An English physicist, mathematician, astronomer, alchemist, inventor, and natural philosopher (1643–1727), who is best known for his work on the laws of motion, optics, gravity, and calculus.

**Newtonian physics:** The branch of mechanics based on Newton's laws of motion.

**Newton's laws of motion:** Physical laws that describe the relationship between the forces acting on a body and its motion due to those forces.

**Nightingale, Florence:** A celebrated English nurse, writer, and statistician (1820–1910) who came to prominence for her work during the Crimean War, transformed nursing by stressing cleanliness, fresh air, and discipline.

**No Child Left Behind Act (NCLB):** United States Act of Congress that supports standards-based education reform based on the premise that setting high standards and establishing measurable goals can improve individual outcomes in education.

**nonexperimental research:** Studies that do not involve a manipulation of the situation, circumstances, or experience of the participants, or any variable that is being measured.

**nonpersister:** A person who does not remain long at any state, purpose, course of action, or the like. Contrast with persister.

**Nostradamus:** A French apothecary and reputed seer (1503–1566) who published collections of prophecies that have since become famous worldwide. Nostradamus has attracted a following that, along with much of the popular press, credits him with predicting many major world events.

**nuclear fission:** The splitting of the nuclei of atoms into two fragments of approximately equal mass, accompanied by conversion of part of the mass into energy.

**nuclear physics:** The branch of physics concerned with the nucleus of the atom.

**nucleic acid:** Any of a group of complex compounds found in all living cells and viruses, composed of purines, pyrimidines, carbohydrates, and phosphoric acid.

**null hypothesis:** A statement that two variables are not related; it contends that there is no effect or no change due to a potential treatment. Contrast with real hypothesis.

**ontology:** The branch of metaphysics concerned with the nature and relations of being; the study of existence.

**Oppenheimer, Robert:** An American theoretical physicist (1904–1967) who taught physics at the University of California, Berkeley. He is often called the "father of the atomic bomb" for his role in the Manhattan Project, the World War II project that developed the first nuclear weapons.

**organ:** A collection of tissues joined in a structural unit to serve a common function.

organelle: A specialized subunit within a cell that has a specific function, and is usually separately enclosed within its own lipid bilayer.

organic chemistry: The branch of chemistry concerned with carbon compounds.

Organization for Economic Co-operation and Development (OECD): An international organization formed to help governments tackle the economic, social, and governance challenges of a globalized economy.

originality: The ability to think and act independently.

osmoregulation: Maintenance of an optimal, constant osmotic pressure in the body of a living organism.

osmosis: The diffusion of a solvent through a semipermeable membrane from a region of low solute concentration to a region of high solute concentration.

Otto, Max: German-born science philosopher (1876–1963) who taught at the University of Wisconsin for 40 years and well known for his views on humanism and pragmatism.

oxygen revolution: The biologically induced appearance of free oxygen ($O_2$) in Earth's atmosphere. This major environmental change happened around 2.4 billion years ago.

Paracelsus: A Swiss alchemist-physician (1493–1541) who rejected the Aristotelian idea that water, fire, earth, and air made up all matter.

parthenogenic birth: A form of reproduction in which the ovum develops into a new individual without fertilization; also known as a virgin birth.

Pavlov, Ivan: Russian physiologist (1849–1936) best known for his study of the conditioned reflex in dogs, whereby he established that many of our human responses are simple mechanical reflexes, starting the behavior psychology field.

peer recognition: Positive acknowledgement of accomplishments from colleagues.

peer review (formal and informal): A process of self-regulation by a profession or a process of evaluation involving qualified individuals within the relevant field.

periodic table: A tabular display of the chemical elements, organized on the basis of their properties. Elements are presented in increasing atomic number.

Perry, William: Well-known educational psychologist (1913–1998) who taught at Harvard Graduate School of Education and studied the cognitive development of students during their college years.

persister: A person who continues steadfastly or firmly in some state, purpose, course of action, or the like, especially in spite of opposition, remonstrance, etc. Contrast with nonpersister.

pet hypothesis: Any theory whose creator favors it more than other theories because he/she wants it to be true, potentially resulting in loss of objectivity.

**philosophy of science:** The study, from a philosophical perspective, of the elements of scientific inquiry.

**pH scale:** A measure of the acidity or alkalinity of a solution; ranges from 0 (acidic) to 14 (basic).

**physical geology:** The branch of geology concerned with understanding the composition of the Earth and the physical changes occurring in it, based on the study of rocks, minerals, and sediments, their structures and formations, and their processes of origin and alteration.

**physics:** The science of energy and matter and how they relate to each other.

**physiology:** The biological study of the functions of living organisms and their parts.

**Piaget, Jean:** A French-speaking Swiss developmental psychologist and philosopher (1896–1980) known for his epistemological studies with children.

**placebo:** A substance containing no medication and prescribed or given to reinforce a patient's expectation to get well.

**plagiarism:** An act or instance of using or closely imitating the language and thoughts of another author without authorization and the representation of that author's work as one's own and not crediting the original author.

**Planck, Max:** A German physicist (1858–1947) who discovered quantum physics, initiating a revolution in natural science and philosophy. He is regarded as the founder of quantum theory, for which he received the Nobel Prize in Physics in 1918.

**plate tectonics:** The study of the structure of Earth's crust and mantle with reference to the theory that the Earth's lithosphere is divided into large rigid blocks (plates) that are floating on semifluid rock and able to interact with each other at their boundaries, and to the associated theories of continental drift and seafloor spreading.

**Plato:** Greek philosopher and perhaps the most influential thinker in the history of Western thought (429–347 B.C.). He was a student of Socrates and a teacher of Aristotle and founded the Academy in Athens where he lectured and taught.

**positive result:** A favorable or concrete outcome or effect characterized by or displaying certainty, acceptance, or affirmation.

**positivism:** A philosophy of science based on the view that in the social as well as natural sciences, data derived from sensory experience, and logical and mathematical treatments of such data, are together the exclusive source of all authentic knowledge.

**positivist perspective:** The use of scientific methods to uncover the laws according to which both physical and human events occur.

**postdoc:** Scholarly research conducted by a person who has recently completed doctoral studies, normally within the first five years; short for postdoctoral.

**post-positivism:** A school of thought which values qualitative over quantitative research, questions the possibility of objectivity, and draws upon the methods of deconstructionism.

**potential energy:** The energy of a body or a system due to the position of the body or the arrangement of the particles of the system.

**prenatal testing:** Testing for diseases or conditions in a fetus or embryo before it is born.

**presentation:** The act of making something publicly available; presenting news or other information by broadcasting or printing it.

*Principia:* Volume of three books by Sir Isaac Newton first published in 1687. It presented a new dynamic mathematical physics, which accounted for the motions of celestial and terrestrial bodies, and includes Newton's laws of motion, which formed the foundation of classical mechanics.

**prokaryote:** An organism that lacks a cell nucleus (karyon), or any other membrane-bound organelles. Contrast with eukaryote.

**prostate specific antigen (PSA) test:** Used to screen for prostate cancer. Measures the amount of prostate-specific antigen in the blood (PSA is released into a man's blood by his prostate gland).

**protein:** A complex, high molecular weight organic compound that consists of amino acids joined by peptide bonds.

**proton:** A subatomic particle found in the nucleus of every atom.

**proximate cause:** An event which is closest to, or immediately responsible for causing, some observed result. Contrast with ultimate cause.

**pseudoscience:** A claim, belief, or practice which is presented as scientific, but does not adhere to a valid scientific method, lacks supporting evidence or plausibility, cannot be reliably tested, or otherwise lacks scientific status.

**psychic prediction:** Knowledge about a future event made by a psychic, someone who is sensitive to forces outside the possibilities defined by natural laws.

**psychogenic:** Originating in the mind or in mental or emotional processes; having a psychological rather than a physiological origin. Used of certain disorders.

**psychological bias:** A pattern of deviation in judgment that occurs in particular situations, leading to perceptual distortion, inaccurate judgment, illogical interpretation, or what is broadly called irrationality.

**Ptolemy:** A second-century Greek scholar (ca. 90–ca. 168) credited with the geocentric model of the universe, the paradigm that places the Earth at the center of the universe.

**qualitative analysis:** The reporting and use of data that are nonnumerical in scope. It usually studies very few subjects or data pieces but looks at those in greater depth than quantitative research.

**quantitative analysis:** The reporting and use of data that are numerical in scope. It is the more traditional scientific analysis system and depends on numbers to find patterns and draw conclusions in the results of a study.

**quark:** Any of a group of elementary particles supposed to be the fundamental units that combine to make up the subatomic particles known as hadrons (baryons, such as neutrons and protons, and mesons).

**quasi-reflective reasoning:** Recognizes that knowledge claims about ill-structured problems contain elements of uncertainty.

**Race to the Top:** Contest created in 2009 to spur innovation and reforms in state and local district K-12 education by awarding points for satisfying certain educational policies, such as performance-based standards for teachers and principals, complying with nationwide standards, promoting charter schools, and computerization.

**radiation:** The transmission of objects or information from a source into a surrounding medium or space. Light, heat, and sound are types of radiation.

**real hypothesis:** A statement that there is a relationship between two variables, or there is a difference between two groups, or there is a difference from a previous or existing standard. Contrast with null hypothesis.

**realism:** Inclination toward literal truth and pragmatism and rejection of the impractical and visionary.

**reasoning:** Use of reason, especially to form conclusions, inferences, or judgments or evidence or arguments used in thinking or argumentation.

**receptor:** A specialized cell or group of nerve endings that responds to sensory stimuli.

**refereed journal article:** A written composition published in a professional publication in which articles or papers are selected by a panel of readers or referees who are experts in the field.

**reflective reasoning:** The capacity to make defensible judgments about complex and controversial issues.

**Reflective Judgment Model:** A model of cognitive development proposed by Patricia King and Karen Kitchener in the 1970s. It describes how people justify their beliefs when faced with vexing problems, characterized by seven distinct but developmentally related sets of assumptions about the process of knowing and how it is acquired.

**relativism:** A theory, especially in ethics, that conceptions of truth and moral values are not absolute but are relative to the persons or groups holding them. Society determines the direction that scientific advancements take.

**Relativist Model:** Theory of intellectual development proposed by William Perry in 1970 that holds that higher levels of critical thinking involve a perception of knowledge and values as contextual and relativistic.

**relativity:** A state of dependence in which the existence or significance of one entity is solely dependent on that of another.

**reliability:** The extent to which an experiment, test, or measuring procedure yields the same results on repeated trials.

**repeatability:** The ability to, under the same conditions as an original study, repeat an investigation and come up with the same conclusions.

**reputation:** A place in public esteem or regard; a good name.

**research problem:** The object or process that needs to be studied.

**residual variation:** The deviation between the values observed and the values that are predicted.

**respiratory acidosis:** A medical condition in which a build-up of carbon dioxide in the blood produces a shift in the body's pH balance and causes the body's system to become more acidic.

**results:** The consequence of a particular action, operation, or course; an outcome.

**rhetoric:** The art or study of using language effectively and persuasively.

**rhetorical argumentation:** A reasoning method that uses observation to argue a scientific point and develop a conclusion to persuade an audience of listeners.

**Sagan, Carl:** An American astronomer, astrophysicist, cosmologist, author, science popularizer, and science communicator in astronomy and natural sciences (1934–1996). He published more than 600 scientific papers and articles and was author, coauthor, or editor of more than 20 books.

**sample group:** A group drawn from a larger population and used to estimate the characteristics of the whole population.

**sampling bias:** When a sample is collected in such a way that some members of the intended population are less likely to be included than others.

**scalar transformations:** Changes made to scales on a graph or a set of data meant to manipulate the appearance of the results.

**scatterplot:** A type of mathematical diagram using Cartesian coordinates to display values for two variables for a set of data.

**scholarly book:** A published text based on original research or experimentation, generally written by a researcher or expert in the field who is affiliated with a college or university.

**science:** A branch of knowledge or study dealing with a body of facts or truths systematically arranged and showing the operation of general laws.

**science, technology, engineering, and mathematics (STEM):** Study of how the world works integrating the four domains of science, technology, engineering, and mathematics.

**scientific literacy:** Comprehension of scientific concepts, processes, values, and ethics, and their relation to technology and society.

**scientific method:** A set of systematic procedures for organized observation and theory-building used to gain scientific knowledge.

**scientific modeling:** The process of generating abstract, conceptual, graphical or mathematical models, an approximation or simulation of a real system that omits all but the most essential variables of the system.

**scientific revolution:** An era associated primarily with the sixteenth and seventeenth centuries during which new ideas and knowledge in physics, astronomy, biology, medicine, and chemistry transformed medieval and ancient views of nature and laid the foundations for modern science.

**scientific statement:** An attempt to explain a scientific happening, a premise.

**second law of thermodynamics:** Relates the motion of molecules to their energy levels. As usable energy is irretrievably lost, disorganization, randomness, and chaos increase.

**sedimentary rock:** Rock that is formed by the deposition of material at the Earth's surface and within bodies of water.

**selfish:** Devoted to or caring only for oneself; concerned primarily with one's own interests, benefits, welfare, etc., regardless of others.

**selfish gene hypothesis:** Theory that evolution occurs through the differential survival of competing genes, increasing the frequency of those alleles whose phenotypic effects successfully promote their own propagation.

**significance level:** A level of error that is acceptable.

**skepticism:** A doubting or questioning attitude or state of mind.

**small diameter nerve fibers:** Peripheral nerve fibers with a low conduction velocity.

**social constructivist perspective:** Encourages the learner to arrive at his or her version of the truth, influenced by his or her background, culture or embedded worldview.

**sociocentricity:** Oriented toward or focused on one's own social group; socially oriented. Contrast with egocentricity.

**Socrates:** Ancient Greek philosopher (470–399 B.C.), and arguably, the founder of the modern scientific method, using questioning and argumentation to bring his students to a greater understanding of truth.

**Socratic method:** A pedagogical technique in which a teacher does not give information directly but instead asks a series of questions, with the result that the student comes either to the desired knowledge by answering the questions or to a deeper awareness of the limits of knowledge.

**species-preserving function:** Behaviors evolved to benefit the group by increasing the survival of related individuals.

**spontaneous generation:** The theory of spontaneous generation held that complex, living organisms may be produced from nonliving matter.

**standard deviation:** A number used to tell how measurements for a group are spread out from the average (mean), or expected value.

**standards for science teachers:** Approved model for science teacher qualifications establishing what teachers are expected to learn and be able to do.

*Streptococcus:* A gram-positive, facultatively anaerobic cocci that occurs in pairs or chains, some of which cause disease.

**string:** A sequence of characters, either as a literal constant or as some kind of variable; the smallest known portion of matter.

**string theory:** Hypothesizes that space-time has more than four dimensions, and that some of the dimensions are exceedingly small and string-like in shape. Elementary particles in string theory are understood as standing waves in such space-time strings, rather than as point-like objects.

**sum of squares between groups:** A function used in statistical analysis to measure variation between the group in a data set.

**sum of squares within groups:** A function used in statistical analysis to measure variation due to differences within individual samples in a data set.

**superstring:** A hypothetical particle consisting of a very short one-dimensional string existing in ten dimensions. It is the elementary particle in a theory of space-time incorporating supersymmetry.

**taxonomy:** A hierarchical system for classifying and identifying organisms.

**tenure:** The status of holding one's position on a permanent basis without periodic contract renewals.

**theory of relativity:** The geometric theory of gravitation published by Albert Einstein in 1916, providing a unified description of gravity as a geometric property of space and time, or space-time.

**thesis statement:** Sentence or two that summarizes the main points and arguments of the author, usually found at the end of the first paragraph of an essay or similar document.

**tissue:** A cellular organizational level intermediate between cells and a complete organism.

**transform fault:** A type of fault in which two tectonic plates slide past one another.

**Trends in International Mathematics and Science Study (TIMSS):** Federal organization that provides reliable and timely data on the mathematics and science achievement of U.S. 4th- and 8th-grade students compared to that of students in other countries.

**Trismegistus, Hermes:** An ancient Egyptian priest, credited as being the first to communicate celestial and divine knowledge to mankind by writing.

**type I error:** Error that incorrectly rejects a hypothesis that is actually correct.

**type II error:**  Error that accepts a hypothesis that is actually wrong.

**typology of argumentation:**  Model of argumentation proposed by Daempfle. A ranking scale that evaluates the strength of scientific arguments, based on King and Kitchener's Reflective Judgment Model.

**ultimate cause:**  The "real" reason something occurred; the underlying cause. Contrast with proximate cause.

**underrepresented groups** (in science):  Insufficiently or inadequately represented within the academic community; results in student isolation and attrition.

**universalism:**  The quality of being inclusive in scope or range, without boundaries.

**valence electron:**  An electron of an atom, located in the outermost (valence) shell of the atom, that can be transferred to or shared with another atom.

**validity:**  Refers to the extent to which a concept, conclusion, or measurement is well-founded and corresponds accurately to the real world.

**van Leeuwenhoek, Anton:**  Dutch scientist (1632–1723) considered to be the first microbiologist. He was the first to observe and describe single-celled organisms, which he originally referred to as *animalcules,* using a handcrafted microscope.

**variance:**  A measure of how far a set of numbers is spread out.

**vasoconstriction:**  The narrowing (constriction) of blood vessels by small muscles in their walls.

**vasodilation:**  Widening of blood vessels that results from relaxation of the muscular walls of the vessels.

**vasodilator:**  Medicine that acts directly on muscles in blood vessel walls to make blood vessels widen (dilate).

**Vesalius, Andreas:**  Flemish anatomist and physician (1514–1564), considered to be the founder of modern human anatomy. His is best known as the author of *De humani corporis fabrica (On the Structure of the Human Body).*

**von Braun, Wehrner:**  German-born rocket scientist, aerospace engineer, and space architect (1912–1977) and one of the leading figures in the development of rocket technology in Nazi Germany during World War II and, subsequently, the United States.

**Watson, James:**  American molecular biologist, geneticist, and zoologist (born 1928) who is credited with the discovery of the structure of the DNA molecule, along with Francis Crick, at Cambridge University in 1953, and was jointly awarded the 1962 Nobel Prize for Physiology or Medicine.

**wave**: A uniformly advancing disturbance in which the parts moved undergo a double oscillation; any wavelike pattern.

**wavelength**: The distance between one peak or crest of a wave of light, heat, or other energy and the next corresponding peak or crest.

**weltanschauung (paradigm)**: A comprehensive conception or image of the universe and of humanity's relation to it.

**whistleblowing**: Occurs when an internal member of an organization reveals misconduct from within to other members of the company or to the public.

# Index

Note: Italicized page locators indicate figures/photos; tables are noted with *t*.